# Intelligent Systems Reference Library

Volume 173

**Series Editors**

Janusz Kacprzyk, Polish Academy of Sciences, Warsaw, Poland

Lakhmi C. Jain, Faculty of Engineering and Information Technology, Centre for Artificial Intelligence, University of Technology, Sydney, NSW, Australia, KES International, Shoreham-by-Sea, UK; Liverpool Hope University, Liverpool, UK

The aim of this series is to publish a Reference Library, including novel advances and developments in all aspects of Intelligent Systems in an easily accessible and well structured form. The series includes reference works, handbooks, compendia, textbooks, well-structured monographs, dictionaries, and encyclopedias. It contains well integrated knowledge and current information in the field of Intelligent Systems. The series covers the theory, applications, and design methods of Intelligent Systems. Virtually all disciplines such as engineering, computer science, avionics, business, e-commerce, environment, healthcare, physics and life science are included. The list of topics spans all the areas of modern intelligent systems such as: Ambient intelligence, Computational intelligence, Social intelligence, Computational neuroscience, Artificial life, Virtual society, Cognitive systems, DNA and immunity-based systems, e-Learning and teaching, Human-centred computing and Machine ethics, Intelligent control, Intelligent data analysis, Knowledge-based paradigms, Knowledge management, Intelligent agents, Intelligent decision making, Intelligent network security, Interactive entertainment, Learning paradigms, Recommender systems, Robotics and Mechatronics including human-machine teaming, Self-organizing and adaptive systems, Soft computing including Neural systems, Fuzzy systems, Evolutionary computing and the Fusion of these paradigms, Perception and Vision, Web intelligence and Multimedia.

** Indexing: The books of this series are submitted to ISI Web of Science, SCOPUS, DBLP and Springerlink.

More information about this series at http://www.springer.com/series/8578

Md Atiqur Rahman Ahad ·
Anindya Das Antar · Masud Ahmed

# IoT Sensor-Based Activity Recognition

Human Activity Recognition

 Springer

Md Atiqur Rahman Ahad
Osaka University
Osaka, Japan

University of Dhaka
Dhaka, Bangladesh

Masud Ahmed
University of Maryland Baltimore County
Maryland, USA

Anindya Das Antar
University of Michigan
Ann Arbor, Michigan, USA

ISSN 1868-4394    ISSN 1868-4408  (electronic)
Intelligent Systems Reference Library
ISBN 978-3-030-51381-8    ISBN 978-3-030-51379-5  (eBook)
https://doi.org/10.1007/978-3-030-51379-5

This Springer imprint is published by the registered company Springer Nature Switzerland AG
The registered company address is: Gewerbestrasse 11, 6330 Cham, Switzerland

*This book is dedicated to—our beloved parents,*
*and our respected teacher (Late) Prof. Zahid Hasan Mahmood, University of Dhaka.*
*M. Ahad would also like to dedicate this book to his lovely daughter Rumaisa Fatima and charming son Zubair Umar!*

# Foreword

Human activity recognition and understanding have been explored in various domains for a long period. Vision-based and sensor-based activity analyses are progressing amazingly with the advent of various IoT sensors and video cameras. The impacts are very high for the present and the future of the world. There are very few genuine books on human activity recognition, and these are mainly in the vision-based field. There is a dire necessity for a comprehensive guideline for the researcher and practitioner in the arena of IoT sensor-based human activity recognition.

This book is filling the huge void by introducing 10 excellent chapters—from the basics of activity recognition to the advanced deep learning related strategies. The book has enriched itself by introducing a number of pragmatic challenges for future research issues. It has a great collection of important references at the end of the book so that readers can go through for further study. The chapters ended with some thought-provoking questions. The book will be very much valuable for now and in the coming years, especially for the undergrad (final year) and Master's course in universities as well as for researchers.

I know Prof. Md Atiqur Rahman Ahad for almost a decade. He has been engaged in research activities and promoting research extensively for a longer period. He has published two books as a single author on vision-based human action recognition in 2011 and 2013. The books are available in Springer. From the records of Springer, it is found that these are well-read and very useful until now. Introducing another book on *IoT Sensor-Based Activity Recognition* is a great move by him. Other co-authors are young and highly-promising researchers as well.

I am very delighted that the book is published by Springer and I am confident that it will get huge circulation in the academic and research communities. I thank Ahad, Antar, and Ahmed for their excellent efforts to produce such a magnificent book.

August 2019                                                            Toshio Fukuda, Fellow IEEE
IEEE President-elect 2019 (President 2020)
Nagoya University, Japan
Meijo University, Japan
Waseda University, Japan
Beijing Institute of Technology, China
Nagoya, Japan

# Comments from Experts

This timely book on sensor-based activity recognition will serve as an excellent overview of the state of the art and a roadmap for what is to come in this area. The authors have nicely presented the relevant problems, approaches, challenges, and opportunities in human activity recognition, identified the most relevant related research, and provided useful summaries and thought-provoking questions for each chapter. In reading this book, readers will gain from their combined expertise and learn important principles for approaching a range of related problems. I highly recommend it.

*Matthew Turk, IEEE Fellow, IAPR Fellow,*
*President, Toyota Technological Institute at Chicago.*

This compilation brings together successfully different aspects and use of IoT sensors in Human Activity Recognition (HAR) with applications in healthcare, elderly people monitoring, fitness tracking, working activity monitoring and more encompassing 150 or so benchmark datasets and dataset repositories for pattern recognition, machine learning, context awareness, and human-centric sensing. In addition to discussing and introducing multiple performance evaluation techniques for use in both video-based and sensor-based (environmental, wearable, and smartphone) HAR, the book highlights deep learning methods to solve the problem of shallow learning using hand-crafted features in conventional pattern recognition approaches.

*Mohammad Karim, IEEE Fellow, OSA Fellow, SPIE Fellow,*
*Provost, Executive Vice Chancellor, University of Massachusetts Dartmouth.*

It is many years now since I first wondered whether people could be recognised by their gait. Then, computers were slow, memory was expensive and accelerometers were enormous. Rolling on 20 years, we find that computers are fast, memory is cheap and the notion that gait is individual to each person is widely accepted. There has also been tremendous progress in sensors and in their analysis and there is now a rich selection of techniques to enable recognition by this. We also now have the IEEE Transactions on Biometrics Behavior and Identity Science. And now we have a book on Sensor-based Activity Recognition. Enjoy!!

Mark Nixon, IET Fellow, IAPR Fellow,
BMVA Distinguished Fellow 2015, University of Southampton.

The book has comprehensive coverage on an increasingly important topic as we start to see more and more wearable devices. The chapters are structured well, with excellent illustrations and activities for use in a graduate or upper-level undergraduate course.

Sudeep Sarkar, AAAS Fellow, IEEE Fellow, AIMBE Fellow, IAPR Fellow,
University of South Florida.

The book is a timely, badly needed, and important treatise on IoT and sensors. Well exposed concepts and algorithms supporting ways of realizing processes of recognition and classification of human activities. Thorough discussions on algorithms, sensing devices and tools as well as benchmark data sets are a genuine asset of this monograph.

Witold Pedrycz, IEEE Fellow,
University of Alberta.

Great work! I especially like the multifaceted approach that includes the design of experiments and the use of available tools, hardware, and methods specifically tailored to activity recognition research.

Kristof Van Laerhoven, Professor,
University of Siegen.

It is one of the most comprehensive, easy-to-read books about the human activity recognition in this modern IoT and Big Data era. Technologies mentioned in this book are up-to-date. Both researchers and practitioners should read this book to grasp its critical ideas and recent advancement.

Atsushi Inoue, Professor,
Eastern Washington University.

This is a cutting-edge collection of theories, technologies, and views on digital human activity recognition. The book is a must-read for students and researchers in the field of IoT sensor-based human activity recognition and its applications. No other book has presented results of this research in a more convincing way.

*Anton Nijholt, Professor*
*University of Twente.*

The elaborate explanations and benefits of utilizing deep learning over conventional pattern recognition approach in the field of human activity recognition with sensor modalities will be beneficial to the research community. Besides, comfortable and straightforward approach of representing benchmark datasets, device information, data collection protocol, and other solution of existing and possible challenges has made this book a must-read.

*Mahbub Hassan, Professor*
*University of New South Wales.*

An excellent book that provides an overview of human activity recognition based on wearable and smartphone sensors.

*Vishal M. Patel, Asst. Professor*
*Johns Hopkins University.*

Small Internet-connected devices with sensors have spread widely to many different application domains. This has left a distinct need for an in-depth look at this emerging field. Developers and researchers who are new to the IoT world will greatly benefit from this comprehensive volume, which helps satisfy that need.

*Walter J. Scheirer, Asst. Professor*
*University of Notre Dame.*

In this book, the authors systematically discuss various components of a sensor-based activity analysis system starting from preprocessing up to performance evaluation. The book provides a clear idea about the purpose and necessity of each component through well-written texts, clean illustrations, and nice data-visualization. The exercises in the 'Think Further' sections at the end of each chapter are well formulated. The book will be a very good first read for any newcomer in the activity analysis domain. On the other hand, the chapters on recent trends and future challenges can provide foods for thought to the experts.

*Upal Mahbub, Senior Engineer*
*Qualcomm Technologies, Inc.*

# Preface

The accelerometer was invented in 1783. Though the initial purpose of using accelerometer was to validate the principles of Newtonian physics, with the advancement of the technology, it has become a popular component in the domain of IoT sensor-based Human Activity Recognition (HAR) in the present days. However, inventions of other sensors (e.g., gyroscope, magnetometer, pressure sensor) enrich this domain a lot. These sensors carry a lot of information, but they are in the form of raw data. There is a plenty of useful information and pattern underneath these raw data, but we need to extract some meaningful inherent information from the data and decipher some patterns. Therefore, we aim to present the tools and techniques, step by step, in this book so that one can acquire the idea of the primary approaches for HAR and progress thereafter. This book cuts through the basic concept of sensor-based human activity classification and demonstrates exactly how and from where to begin with, if someone is a newcomer in this research arena.

We can divide this book into three parts:

- At the beginning, we discuss different approaches of HAR, different types of filters for removing noises from the raw data, various parameters of those filters, and their effects, and different types of windowing techniques. Following these aspects, we amass several essential features in the time, frequency and other domains that are conventionally explored by the researchers. We enlighten on how to select important features, deduce surplus features, prepare them for the classification, and classify various activities using those features.
- In the next part of this book, we scrutinize distinctive issues and factors that we have to consider while designing a new dataset related to sensor-based human activity analysis. Moreover, we introduce different tools and applications that

are exploited for data collection. In this book, we summarize about 150
benchmark datasets and categorize them with some important features.

- In the final section of the book, we present more information about classification
  and evaluation strategies. We bestow on deep learning concepts and how this
  extremely flourishing domain can solve the problems of classification faced by
  the conventional statistical classifiers. Finally, we demonstrate a number of
  future challenges that one can ponder on for further developments.

We introduce some thought-provoking questions at the end of each chapter. We
strongly recommend that a reader explore the entire book to get a comprehensive
exploration of sensor-based human activity recognition and analysis.

We are delighted to present this book for the students of upper level of undergrad
and postgraduate, as well as, researchers in academia and industry—in the domain
of IoT, sensor, HAR, healthcare, machine learning and related fields. We firmly
anticipate that this book will be a genuine companion on the *IoT Sensor-based
Human Activity Recognition* journey. We have been engaged ourselves for this
book for about three years' period, and we kept on polishing the book for a longer
period to enrich the content and the clarity.

We are indebted to **Toshio Fukuda**, *Fellow IEEE, IEEE President 2020* for his
time to write the 'Foreword' of this book. We would like to offer our sincere
gratitude to a panel of great researchers who poured their valuable time and
comments to enrich the book. We would like to mention them with our sincerest
gratefulness: **Matthew Turk** (*IEEE Fellow, IAPR Fellow*, President, Toyota
Technological Institute at Chicago), **Mohammad Karim** (*IEEE Fellow, OSA
Fellow, SPIE Fellow*, Provost, Executive Vice Chancellor, University of
Massachusetts Dartmouth), **Sudeep Sarkar** (*AAAS Fellow, IEEE Fellow, AIMBE
Fellow, IAPR Fellow*, Professor, University of South Florida), **Witold Pedrycz**
(*IEEE Fellow*, Canada Research Chair, University of Alberta), **Mark Nixon** (*IET
Fellow, IAPR Fellow, BMVA Distinguished Fellow 2015*, Professor, University of
Southampton), **Diane J. Cook** (*IEEE Fellow, FTRA Fellow, NAI Fellow*,
Huie-Rogers Chair Professor, Washington State University), **Kenichi Kanatani**
(*IEEE Fellow*, Professor Emeritus, Okayama University), **Atsushi Inoue**
(Professor, Eastern Washington University), **Kristof Van Laerhoven** (Professor,
University of Siegen), **Anton Nijholt** (Professor, University of Twente), **Mahbub
Hassan** (Professor, University of New South Wales), **Vishal M. Patel** (Asst.
Professor, Johns Hopkins University), **Walter J. Scheirer** (Asst. Professor,
University of Notre Dame.), and **Upal Mahbub** (Senior Engineer, Qualcomm
Technologies, Inc.).

We want to thank everyone who encouraged and assisted us to accomplish this
book. Finally, we are grateful to the Springer and Prof. Lakhmi Jain for their

endorsements and publishing the book. We will be delighted to have your valuable feedback on this book. Enjoy the book!

Osaka, Japan

Md Atiqur Rahman Ahad, SMIEEE
Associate Professor, Osaka University
Professor, University of Dhaka
atiqahad@du.ac.bd
http://ahadVisionLab.com
http://cennser.org/ICIEV
http://cennser.org/IVPR
http://cennser.org/IJCVSP
https://abc-research.github.io/

Michigan, USA

Anindya Das Antar
adantar@umich.edu

Maryland, USA

Masud Ahmed
mahmed10@umbc.edu

# Acknowledgments

We would like to specially thank the following researchers for their time to look into the book and various comments to improve:

**Toshio Fukuda** (*IEEE Fellow*, **IEEE President 2020**, Nagoya University, Meijo University, Waseda University, Beijing Institute of Technology)

**Matthew Turk** (*IEEE Fellow, IAPR Fellow*, President, Toyota Technological Institute at Chicago)

**Mohammad Karim** (*IEEE Fellow, OSA Fellow, SPIE Fellow*, Provost, Executive Vice Chancellor, University of Massachusetts Dartmouth)

**Sudeep Sarkar** (*AAAS Fellow, IEEE Fellow, AIMBE Fellow, IAPR Fellow*, Professor, University of South Florida)

**Witold Pedrycz** (*IEEE Fellow*, Canada Research Chair, University of Alberta)

**Mark Nixon** (*IET Fellow, IAPR Fellow, BMVA Distinguished Fellow 2015*, Professor, University of Southampton)

**Diane J. Cook** (*IEEE Fellow, FTRA Fellow, NAI Fellow*, Huie-Rogers Chair Professor, Washington State University)

**Kenichi Kanatani** (*IEEE Fellow*, Professor Emeritus, Okayama University)

**Atsushi Inoue** (Professor, Eastern Washington University)

**Kristof Van Laerhoven** (Professor, University of Siegen)

**Anton Nijholt** (Professor, University of Twente)

**Mahbub Hassan** (Professor, University of New South Wales)

**Vishal M. Patel** (Asst. Professor, Johns Hopkins University)

**Walter J. Scheirer** (Asst. Professor, University of Notre Dame.)

**Upal Mahbub** (Senior Engineer, Qualcomm Technologies, Inc.)

# Contents

# Acronyms

AUC      Area Under The Curve
AUROC      Area Under the Receiver Operating Characteristics
BA      Body Acceleration
BCR      Balanced Classification Rate
BER      Balanced Error Rate
CART      Classification and Regression Trees
CFS      Correlation-based Feature Selection
CNN      Convolutional Neural Network
DBN      Deep Belief Network
DCT      Discrete Cosine Transform
DNN      Deep Neural Network
DT      Decision Tree
GA      Gravitational Acceleration
HAPT      Human Activity with Postural Transitions
HAR      Human Activity Recognition
HASC      Human Activity Sensing Consortium
ICA      Independent Component Analysis
IQR      Interquartile Range
KNN      K-Nearest Neighbour
LDA      Linear Discriminant Analysis
LR      Logistic Regression
LSTM      Long Short-Term Memory
MAD      Median Absolute Deviation
MRMR      Maximum Relevance and Minimum Redundancy
PCA      Principal Component Analysis
PR      Pattern Recognition
PT      Postural Transition
RBM      Restricted Boltzmann Machine
RMS      Root Mean Square
RnF      Random Forest

| RNN | Recurrent Neural Network |
|-----|--------------------------|
| ROC | Receiver Operating Characteristic |
| SAE | Stacked Autoencoder |
| STFT | Short-Time Fourier Transforms |
| SVM | Support Vector Machine |
| t-SNE | t-Distributed Stochastic Neighbor Embedding |
| UMAP | Uniform Manifold Approximation and Projection |

# List of Figures

# List of Tables

# Chapter 1
# Introduction on Sensor-Based Human Activity Analysis: Background

**Abstract** The constant growth of sensor-based systems and technologies for the detection of human activities has made notable progress in the field of human-computer interaction. The continuation of Internet connectivity into daily objects and physical devices has made it possible for the researchers to use IoT sensors for healthcare, elderly people monitoring, fitness tracking, working activity monitoring, and so on. The prominent application fields of sensor-based activity monitoring systems are many, but not limited to, pattern recognition, machine learning, context awareness, and human-centric sensing. If a salient investigation is performed on this topic by fellow researchers, this can create a vital turn in the way of interaction among people and mobile devices. In this book, we have bestowed a comprehensive survey showing the various aspects of human activity recognition based on wearable, environmental, and smartphone sensors. After discussing the background, numerous factors have been analyzed for the data pre-processing part regarding noise filtering and segmentation methods. The list of sensing devices, sensors, and application tools listed in this book can be used for the activity data collection efficiently. Moreover, a detailed analysis of more than 150 benchmark datasets and dataset repositories in this book includes information about sensors, attributes, activity classes, etc. These datasets sum up several types of sensor-based daily activities, medical activities, fitness activities, transportation activities, device usage, fall detection, and hand gesture data. In addition to these, we have shown the feature extraction and classical machine learning methods in detail. Moreover, the overview of different types of classification problems has been given along with the discussion on several performance evaluation techniques showing their advantages and limitations. Furthermore, we have also discussed the importance of deep learning methods to solve the problem of shallow learning using hand-crafted features in conventional pattern recognition approaches. Finally, we have presented a summary of activity recognition methods focused on recent works in several benchmark datasets, and mentioned some future challenges regarding data collection, design issues, and other prospects.

© Springer Nature Switzerland AG 2021                                          1
M. A. R. Ahad et al., *IoT Sensor-Based Activity Recognition*, Intelligent Systems
Reference Library 173, https://doi.org/10.1007/978-3-030-51379-5_1

## 1.1  Introduction

Human activity recognition (HAR) has been one of the most dominant and influential research subjects in various fields over the last few decades, including mobile computing [3, 4], machine learning, context-aware computing [5, 6], security based on surveillance [7–10], energy expenditure estimation [11–13] and analysis [14, 15], fall detection [16, 17], fall risk assessment [18], motor rehabilitation [19], transportation activity monitoring [20, 21], age and gender estimation [22, 23], cardiac monitoring [24], and ambient assistive living [25–27]. The goal of activity recognition is to identify the actions executed by a people delivered a set of inspections of itself and the enclosing environment. HAR seeks to understand people's everyday activities by examining insights gathered from individuals and their surrounding living environments. This knowledge is gathered from various IoT sensors embedded in smartphones, wearable devices and home settings [28]. Cameras and video-based devices are also used in the domain of computer vision (e.g., as RGB frames, Depth maps, and skeleton joints) to record day-to-day human activities for automated recognition systems [29–32].

Comprehensive growth of low-power, low-cost, and miniaturized but high-capacity- or density-based sensors, along with the flourished wired and wireless communication networks [33–35], the sensor-based domains have emerged tremendously. Therefore, interactions with devices and human beings have progressed a lot. Sensors become integral parts of our daily lives in most of the sectors. Besides, in present days due to the access of various sensors in smartphones, there has been a transformation towards portable mobile phones in current years from dedicated wearable motion sensors. We have shown some basic embedded smartphone sensors in Fig. 1.1, which can be utilized in activity recognition.

Though there are numerous applications, the extensive goal of most of the research works in human activity identification is the remote monitoring of the regular activities of people, especially of pregnant women, elderly people, and hospital patients. It

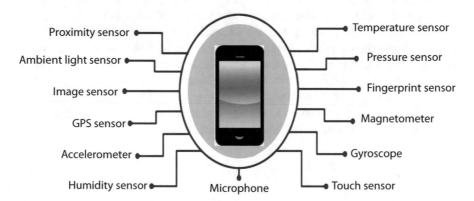

**Fig. 1.1** Basic embedded smartphone sensors to monitor human activities

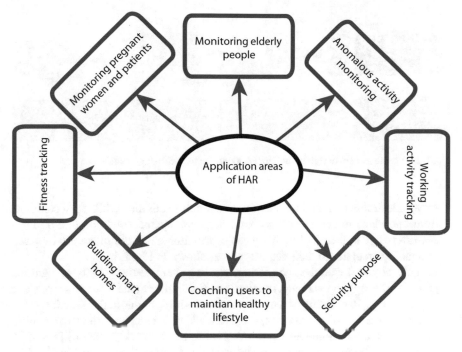

**Fig. 1.2** Application areas of human activity recognition (HAR)

is targeted to have 24hr monitoring or evaluation to provide them ubiquitous health and well-being supervisions [36]. According to the current population benchmark, the world population is increasing expeditiously. If we look at the World Population Aging Report 2019, we will find that number of persons aged 65 or over in 2019 was 702.9 million all over the world. It is expected to be 1548.9 million within 2050 (120% change within 30 years). Moreover, in 2019, these numbers were 200.4 million in Europe and Northern America, 119 million in Central and Southern Asia, 260.6 million in Eastern and South-Eastern Asia, 4.8 million in Australia and New Zealand, and 56.4 million in Latin America and the Caribbean [37].

Besides this, the condition worsens by the rapid increase of nuclear families and abroad or big cities from rural areas or remote islands going students, job-seekers who have to leave their older parents alone at home. This condition desires more welfare services and provisions to the elderly support system and by tracking their daily activities, we can remotely monitor them. The same process can be useful for the supervision and fast response to pregnant women and patients suffering from infirmity or persistent disorders such as Parkinson's disease (PD), autism, and visual impairments [38]. Likewise, patients suffering from dementia, flu, and insanity can be monitored to detect anomalous activities and thereby inhibit unwanted consequences [39, 40]. We have summarized the application areas of human activity recognition (HAR) in Fig. 1.2. These are:

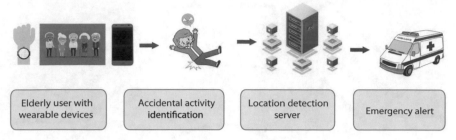

**Fig. 1.3** A basic system design to monitor accidental activities using wearable sensors

- Any undesirable change in elderly people's daily activities while they commute from one place to another or any kind of unanticipated condition that brings a drastic change in their daily activity can be monitored and their lives can be saved by sending an alarm to their nearest ones as shown in Fig. 1.3.
- In daily physical exercise monitoring, we can also use the activity recognition process, like jogging, walking, running, etc. Furthermore, the recognition of static activities, e.g., like, sitting, standing, etc. along with dynamic activities with postural transitions like sitting to stand, stand to walk, etc. It can be assistive to monitor the workers, their fatigue levels, working rates, etc. in working sectors [41–46].
- Such assessments and programs can be essential if people are to maintain a healthy lifestyle by recommending minute behavioral corrections. For instance, we can inspire the users to avoid elevators and use stairs regularly, or we can alarm them to stand after an elongated span of sitting at workplaces.
- On-body sensing devices can also help to detect drug-seeking behavior by exploiting respiratory, cardiac and other vital signals [47].
- In strategic circumstances, soldiers are expected to obtain detailed information about their actions, health conditions, and locations for protection and safety purposes. This knowledge can be particularly useful in combat and training conditions when it comes to decision taking.
- Smart home [48, 49] for daily activity monitoring or activities of daily living (ADL) can provide external sensing to monitor moderately complicated or complex daily activities (e.g., cooking, working on a computer, using utensils, brushing teeth, using toilet, eating lunch, etc.). This concept is based on numerous sensors merged in different locations of home and home appliances, which users are assumed to interact with (e.g., bed, faucet, stove, washing machine, oven, and locker).
- Camera-based systems have been extensively explored for video-based surveillance (e.g., intervention disclosure) and interactive purposes. Kinect game console [50], which has been developed by Microsoft has been used for depth maps and skeleton-based applications.

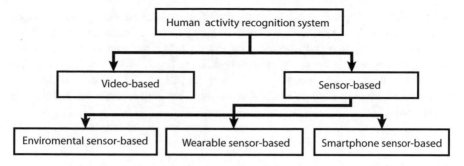

**Fig. 1.4** Approaches for human activity recognition

## 1.2 Approaches to Human Activity Recognition

To achieve the goal of recognizing human activity, we need systems with sensing capabilities that recognize activity. For this reason two approaches are mainly employed:

- Video-based action or activity recognition and
- Sensor-based human activity recognition.

Sensor-based human activity recognition approaches are split into the following three categories:

- Environmental sensor-based activity recognition,
- Wearable sensor-based activity recognition, and
- Smartphone sensor-based systems, as shown in Fig. 1.4.

## 1.3 Comparison of Different Approaches

A basic comparison of different approaches of human activity recognition techniques and procedures have been analyzed in detail mentioning different modalities in this section. The advantages and disadvantages of these approaches have also been discussed in detail.

### 1.3.1 Video-Based Activity Recognition

Video-based approaches often work well indoor environment, especially for depth map-based methods. However, these video-based methods do not succeed to achieve comparable results in outdoor or in real-life situations. Some of the core challenges are: variable lighting conditions—hence, variations in illuminations, cluttered environments, and high diversification of actions that take place in natural surroundings and so on [51, 52]. In a multi-person scenario, an individual's action understanding and segmenting each subject are extremely difficult tasks. Moreover, this system is expensive, infrastructure dependent, and hampers privacy. So, we have focused on sensor-based human activity recognition, as this field is growing more attention in recent days.

### 1.3.2   Environmental Sensor-Based Activity Recognition

Environmental sensor-based systems consist of ambient sensors, which are disseminated throughout the subject's sustenance environment. This kind of systems passively monitors their occupants, thus the users do not need to operate manually. A variety of sensors can be used for this purpose to monitor a vast amount of parameters. Ambient sensors are easy to install and they have fewer limitations in terms of size, weight, and power than other types of sensors, which simplifies the system design. However, the main problem regarding these sensors is infrastructure dependency and failure of monitoring subjects outside the home environment. Also, they manifest complications to differentiate between the subject to be monitored and the neighborhood in the residence.

### 1.3.3   Wearable Sensor-Based Activity Recognition

Wearable sensor-based systems are intended to be attached to a part or multiple locations of the human body during daily activities. These are mostly done for the continuous measurement of activity records, physiological and biomechanical data collections, irrespective of subject's locations [53–55]. Over an extended period, wearable sensors are suitably adapted to gather data on regular physical activity patterns as they can be integrated into jewelry [56], clothing [57], earbud [58], or worn as wearable devices like a smartwatch, wristband, gloves, etc. Wearable sensors can capture some physiological data and regulate some parameters, which may not be measurable by using camera devices or ambient sensors. The reason is that an wearable sensor remains attached to the monitoring subject and it is invariant of any environment (usually) and any infrastructure.

For example, wearable sensors are extensively used for monitoring of activities like sleep monitoring (though there are different types of sensors or systems for sleep monitoring and performances vary significantly from one to another). Moreover, these sensors are cheaper than other types of sensing methods like audio, video, EEG, ECG, etc. and they have no typical privacy issues, unlike video sensors. On-body sensors are proficient in measuring a variety of body signals including motion, physiological signals, location, etc. [59]. This capability not only enhances patient monitoring task but also provides portable and remote supervision of elderly people. But unfortunately, two of the main constraints of on-body sensors are the discomfort while wearing sensors by a patient or elderly person, and various vulnerabilities. Battery-life is another important concern. Moreover, it will be harder to make the wearable system stable while performing daily activities, which creates fluctuations in accelerometer and gyroscope data.

### 1.3.4 Smartphone Sensor-Based Activity Recognition

Nowadays, smartphones arise from the combination of new advantages and features (e.g., Internet access, location-based services, gaming, and multi-sensing capabilities) that complement the conventional telephone service. Smartphones can play a critical role in exploring novel solutions for retrieving data directly from the consumers. One of the benefits of today's mobile developments is that they combine inertial sensors such as accelerometers, gyroscopes, magnetometers and so on that can be used for work on detection of human activities.

However, a smartphone may run a number of active applications and sensors, and data transfer. These can quickly commit for energy loss of the smartphone's battery. Besides, it requires various floating-point operations to be carried out per second if we want to exploit different models for human activity recognition on smartphones. However, this may not be an issue but this could lead to a problem with quick battery discharge, which can affect user's attention. However, fixed-point arithmetic based system reforming the multiclass conventional Support Vector Machine (SVM) can solve the problem [60]. Another positive side of activity data collection using internal smartphone sensor is flexibility. We can easily engage lots of users with smartphones for the data collection process and access the data using a mobile application. For all these reasons, smartphone sensor-based activity recognition is an open research area considering the challenges of less battery usage, sensible way of data collection, behavior pattern analysis, etc.

However, recognizing daily activities will be a great difficulty for an autonomous system and a vast number of other sensory data will be required. Because of the deficiency of good review of the structural basis of activity recognition and lacking relevant benchmark dataset information, most of the researchers find it difficult to research in this field. There are several good survey papers like [9, 61, 62] on vision-based activity recognition technique but there is a lacking of comprehensive works on sensor-based activity recognition, which is needed for the community.

Therefore, in this book, we have summed up more than 150 benchmark datasets on wearable, smartphone, and environmental sensor-based human activity recognition along with fall detection [17] and transportation activities.

## 1.4   Outline of the Book

The basic outline of this book have been presented below that covers a background analysis of human activity recognition research with sensor modality, methods of data preprocessing, feature analysis, conventional pattern recognition approaches used in previous research works, existing challenges, information about available sensing devices, analysis of benchmark datasets, overview of classification problems, performance evaluation techniques, evolution of deep learning-based approaches, and future challenges.

- In this chapter includes an introduction to the definitions, concepts and principles of human activity recognition systems along with the historical perspective, where intentions and motivation of research in this field have been clarified.
- The importance of human activity recognition (HAR), its application in numerous fields have been also discussed. Besides, several ways for pre-processing and segmentation steps of raw-sensor data have been discussed in Chap. 2.
- Chapter 3 provides feature extraction, selection and classification procedures.
- We have presented some challenges in human activity recognition regarding various determinants including the number of activity classes, the types of classes and their relationships, choice of sensors, energy consumption, data collection protocols, etc. in Chap. 4.
- Chapter 5 provides a brief description of a number of sensing devices, systems and application tools that can be used in making a dataset on activity recognition.
- We have provided an in-depth presentation for more than 150 benchmark datasets in Chap. 6. We have discussed some required information of those datasets under various categories and have added dataset reference, availability, name, and number of sensors, devices being used, number of subjects, number of activity classes, etc.
- In Chapter 7, we have given an overview of classification problems (binary classification, multi-class classification, etc.) using proper explanation with examples.
- The performance measure and evaluation matrices (confusion matrix, ROC curve, accuracy, precision, recall, etc.) to justify the performance of classifiers have been discussed in detail in Chap. 8.
- Chapter 9 shows the evaluation of deep models to solve the poor generalization and shallow learning problem of conventional pattern recognition approach.
- We have introduced various future challenges in the field of human activity recognition (HAR) in Chap. 10. Finally, this chapter concludes and provides directions for future research.

## 1.5   Conclusion

In this chapter, we have presented sensor-based activity recognition related various aspects after analyzing the background. There are different approaches on sensor-based activity recognition and some glimpses are presented in this chapter. It is to be noted that video-based action/activity recognition strategies are different from basic wearable sensor-based activity recognition. Variations among different parameters are widely varied in both categories. The advantages and disadvantages of using different sensor modalities for activity recognition can be found in this chapter.

## 1.6 Think Further

1. What are the application areas of Human Activity Recognition (HAR)?
2. What are the basic approaches for HAR?
3. Which type of sensors can be used in HAR?
4. What are the drawbacks of vision-based human activity recognition approach?
5. What are the benefits of sensor-based human activity recognition?
6. What are the major settings and differences among environmental, wearable, and smartphone sensor-based HAR?
7. How can we solve user comfort issue in the case of wearable system?
8. What are the application areas of environmental sensor-based HAR system?
9. What are the importance of remote monitoring of elderly people?
10. What are the motivations behind sensor-based HAR research?

## References

1. Ahmed, M., Das Antar A., Ahad, M.A.R.: An approach to classify human activities in real-time from smartphone sensor data. In: 2019 Joint 8th International Conference on Informatics, Electronics Vision (ICIEV) and 2019 3rd International Conference on Imaging, Vision Pattern Recognition (icIVPR)
2. Antar, A.D., Ahmed, M., Ahad, M.A.R.: Challenges in sensor-based human activity recognition and a comparative analysis of benchmark datasets: a review. In: 2019 Joint 8th International Conference on Informatics, Electronics & Vision (ICIEV) and 2019 3rd International Conference on Imaging, Vision & Pattern Recognition (icIVPR), pp. 134–139. IEEE, Cheney, WA (2019)
3. Weiser, M.: The computer for the 21st century. Sci. Am. **265**(3), 94–105 (1991)
4. Choudhury, T., Consolvo, S., Harrison, B., Hightower, J., LaMarca, A., LeGrand, L., Rahimi, A., Rea, A., Bordello, G., Hemingway, B., et al.: The mobile sensing platform: an embedded activity recognition system. IEEE Pervas. Comput. **7**(2), 32–41 (2008)
5. Laerhoven, V.K., Aidoo, K.: Teaching context to applications. Pers. Ubiquitous Comput. **5**(1), 46–49 (2001)
6. Wren, C.R., Tapia, E.M.: Toward scalable activity recognition for sensor networks. In: International Symposium on Location-and Context-Awareness, pp. 168–185. Springer, Berlin (2006)
7. Aggarwal, J.K., Cai, Q.: Human motion analysis: a review. Comput. Vis. Image Understand. **73**(3), 428–440 (1999)
8. Cedras, C., Shah, M.: Motion-based recognition a survey. Image Vis. Comput. **13**(2), 129–155 (1995)

9. Poppe, R.: A survey on vision-based human action recognition. Image Vis. Comput. **28**(6), 976–990 (2010)
10. Moeslund, T.B., Hilton, A., Krüger, V.: A survey of advances in vision-based human motion capture and analysis. Comput. Vis. Image Understand. **104**(2–3), 90–126 (2006)
11. Bouten, C.V.C., Sauren, A.A.H.J., Verduin, M., Janssen, J.D.: Effects of placement and orientation of body-fixed accelerometers on the assessment of energy expenditure during walking. Med. Biol. Eng. Comput. **35**(1), 50–56 (1997)
12. Swartz, A.M., Strath, S.J., Bassett, D.R., O'brien, W.L., King, G.A., Ainsworth, B.E.: Estimation of energy expenditure using CSA accelerometers at hip and wrist sites. Med. Sci. Sports Exerc. **32**(9), S450–S456 (2000)
13. Crouter, S.E., Clowers, K.G., Bassett, D.R. Jr.: A novel method for using accelerometer data to predict energy expenditure. J. Appl. Physiol. **100**(4), 1324–1331 (2006)
14. Mayagoitia, R.E., Lötters, J.C., Veltink, P.H., Hermens, H.: Standing balance evaluation using a triaxial accelerometer. Gait Posture **16**(1), 55–59 (2002)
15. Moe-Nilssen, R., Helbostad, J.L.: Trunk accelerometry as a measure of balance control during quiet standing. Gait Posture **16**(1), 60–68 (2002)
16. Pannurat, N., Theekakul, P., Thiemjarus, S., Nantajeewarawat, E.: Toward real-time accurate fall/fall recovery detection system by incorporating activity information. In: Proceedings of 2012 IEEE-EMBS International Conference on Biomedical and Health Informatics, pp. 196–199. IEEE, Hong Kong (2012)
17. Islam, M.Z., Serikawa, S., Islam, Z.Z., Tazwar, S.M., Ahad, M.A.R.: Automatic fall detection system of unsupervised elderly people using smartphone. In: Annual Conference on Artificial Intelligence. IEEE, Hawaii (2017)
18. King, R.C., Atallah, L., Wong, C., Miskelly, F., Yang, G.-Z.: Elderly risk assessment of falls with bsn. In: 2010 International Conference on Body Sensor Networks, pp. 30–35. IEEE, Singapore (2010)
19. Fortino, G., Gravina, R.: Rehab-aaservice: a cloud-based motor rehabilitation digital assistant. In: Proceedings of the 8th International Conference on Pervasive Computing Technologies for Healthcare, pp. 305–308. ICST (Institute for Computer Sciences, Social-Informatics and Telecommunications Engineering (2014)
20. Bi, C., Xing, G.: Ramt: Real-time attitude and motion tracking for mobile devices in moving vehicle. Proc. ACM Interact. Mob. Wearable Ubiquitous Technol. **3**(2), 38 (2019)
21. Fang, Z., Yang, Y., Wang, S., Fu, B., Song, Z., Zhang, F., Zhang, D.: Mac: Measuring the impacts of anomalies on travel time of multiple transportation systems. Proc. ACM Interact. Mob. Wearable Ubiquitous Technol. **3**(2), 42 (2019)
22. Antar, A.D., Ahmed, M., Hossain, T., Muramatsu, D., Makihara, Y., Inoue, S., Yagi, Y., Ahad, M.A.R., Ngo, T.T.: Wearable sensor-based gait analysis for age and gender estimation (2020)
23. Ngo, T.T., Ahad, M.A.R., Antar, A.D., Ahmed, M., Muramatsu, D., Makihara, Y., Yagi, Y., Inoue, S., Hossain, T., Hattori, Y.: Ou-isir wearable sensor-based gait challenge: age and gender. In: Proceedings of the 12th IAPR International Conference on Biometrics, ICB (2019)
24. Covello, R., Fortino, G., Gravina, R., Aguilar, A., Breslin, J.G.: Novel method and real-time system for detecting the cardiac defense response based on the ECG. In: 2013 IEEE International Symposium on Medical Measurements and Applications (MeMeA), pp. 53–57. IEEE, Crete (2013)
25. Philipose, M., Fishkin, K.P., Perkowitz, M., Patterson, D.J., Fox, D., Kautz, H., Hahnel, D.: Inferring activities from interactions with objects. IEEE Pervas. Comput. **3**(4), 50–57 (2004)
26. Cook, D.J., Schmitter-Edgecombe, M.: Assessing the quality of activities in a smart environment. Methods Inf. Med. **48**(5), 480 (2009)
27. Ravi, N., Dandekar, N., Mysore, P., Littman, M.L.: Activity recognition from accelerometer data. In: Aaai, vol. 5, pp. 1541–1546 (2005)
28. Garg, R., Moreno, C.: Understanding motivators, constraints, and practices of sharing internet of things. Proc. ACM Interact. Mob. Wearable Ubiquitous Technol. **3**(2), 44 (2019)
29. Turaga, P., Chellappa, R., Subrahmanian, V.S., Udrea, O.: Machine recognition of human activities: a survey. IEEE Trans. Circuits Syst. Video Technol. **18**(11), 1473–1488 (2008)

30. Ahad, M.A.R.: Motion History Images for Action Recognition and Understanding. Springer Science & Business Media, Berlin (2012)
31. Ahad, M.A.R.: Computer Vision and Action Recognition: A Guide for Image Processing and Computer Vision Community for Action Understanding, vol. 5. Springer Science & Business Media, Berlin (2011)
32. Antar, A.D., Ahad, M.A.R., Shahid, O.: Vision-based action understanding for assistive healthcare: a short review. In: IEEE CVPR Workshop (2019)
33. Pantelopoulos, A., Bourbakis, N.G.: A survey on wearable sensor-based systems for health monitoring and prognosis. IEEE Trans. Syst. Man Cybernet. Part C (Appl. Rev.) **40**(1), 1–12 (2010)
34. Alemdar, H., Ersoy, C.: Wireless sensor networks for healthcare: a survey. Comput. Netw. **54**(15), 2688–2710 (2010)
35. Ding, D., Cooper, R.A., Pasquina, P.F., Fici-Pasquina, L.: Sensor technology for smart homes. Maturitas **69**(2), 131–136 (2011)
36. Choi, W., Park, S., Kim, D., Lim, Y.-K., Lee, U.: Multi-stage receptivity model for mobile just-in-time health intervention. Proc. ACM Interact. Mob. Wearable Ubiquitous Technol. **3**(2), 39 (2019)
37. World Population Ageing.: Department of economic and social affairs population division (2019). https://www.un.org/en/development/desa/population/publications/pdf/ageing/WorldPopulationAgeing2019-Highlights.pdf. Accessed 25 March 2020
38. Ward, J.A., Richardson, D., Orgs, G., Hunter, K., Hamilton, A.: Sensing interpersonal synchrony between actors and autistic children in theatre using wrist-worn accelerometers. In: Proceedings of the 2018 ACM International Symposium on Wearable Computers, pp. 148–155. ACM, New York (2018)
39. Yin, J., Yang, Q., Pan, J.J.: Sensor-based abnormal human-activity detection. IEEE Trans. Knowl. Data Eng. **20**(8), 1082–1090 (2008)
40. Dong, W., Guan, T., Lepri, B., Qiao, C.: Pocketcare: Tracking the flu with mobile phones using partial observations of proximity and symptoms. Proc. ACM Interact. Mob. Wearable Ubiquitous Technol. **3**(2), 41 (2019)
41. Islam, N.R.A., Ahad, M.A.R.: A study on tiredness assessment by using eye blink detection. pp. 209–214 (2019)
42. Mirjafari, S., Masaba, K., Grover, T., Wang, W., Audia, P., Campbell, A.T., Chawla, N.V., Swain, V.D., Choudhury, M.D., Dey, A.K., et al.: Differentiating higher and lower job performers in the workplace using mobile sensing. Proc. ACM Interact. Mob. Wearable Ubiquitous Technol. **3**(2), 37 (2019)
43. Noman, M.T.B., Hussein, M.A., Ahad, M.A.R.: A study on tiredness measurement using computer vision. pp. 110–117 (2019)
44. Syeda, U.H., Zafar, Z., Islam, Z.Z., Tazwar, S.M., Rasna, M.J., Kise, K., Ahad, M.A.R.: Visual face scanning and emotion perception analysis between autistic and typically developing children. In: ACM UbiComp Workshop on Mental Health and Well-being: Sensing and Intervention. ACM, New York (2017)
45. Sami, M., Irtija, N., Ahad, M.A.R.: Fatigue detection using facial landmarks. In: 4th International Symposium on Affective Science and Engineering, and the 29th Modern Artificial Intelligence and Cognitive Science Conference (ISASE-MAICS) (2018)
46. Noman, M.T.B., Ahad, M.A.R.: Mobile-based eye-blink detection performance analysis on android platform (2018)
47. Gullapalli, B.T., Natarajan, A., Angarita, G.A., Malison, R.T., Ganesan, D., Rahman, T.: On-body sensing of cocaine craving, euphoria and drug-seeking behavior using cardiac and respiratory signals. In: Proceedings of the ACM on Interactive, Mobile, Wearable and Ubiquitous Technologies, **3**(2), 46 (2019)
48. Van Kasteren, T.L.M., Englebienne, G., Kröse, B.J.A.: An activity monitoring system for elderly care using generative and discriminative models. Pers. Ubiquitous Comput. **14**(6), 489–498 (2010)

49. Andrei, T., Xin, H., Jit, B., Chris, N., Liming, C., Guido, P.: Comparison of fusion methods based on dst and dbn in human activity recognition. J. Control Theory Appl. **9**(1), 18–27 (2011)
50. Shotton, J., Fitzgibbon, A., Cook, M., Sharp, T., Finocchio, M., Moore, R., Kipman, A., Blake, A.: Real-time human pose recognition in parts from single depth images. In: Computer Vision and Pattern Recognition (CVPR), 2011 IEEE Conference on, pp. 1297–1304. IEEE, Washington, DC (2011)
51. Tapia, E.M., Intille, S.S., Larson, K.:. Activity recognition in the home using simple and ubiquitous sensors. In: International Conference on Pervasive Computing, pp. 158–175. Springer, Berlin (2004)
52. Ahad, A.R. Md.: Vision and sensor based human activity recognition: Challenges ahead (2020)
53. Najafi, B., Aminian, K., Paraschiv-Ionescu, A., Loew, F., Bula, C.J., Robert, P.:. Ambulatory system for human motion analysis using a kinematic sensor: monitoring of daily physical activity in the elderly. IEEE Trans. Biomed. Eng. **50**(6), 711–723 (2003)
54. Costa, J., Guimbretière, F., Jung, M.F., Choudhury, T.: Boostmeup: Improving cognitive performance in the moment by unobtrusively regulating emotions with a smartwatch. Proc. ACM Interact. Mob. Wearable Ubiquitous Technol. **3**(2), 40 (2019)
55. Elkader, S.A., Barlow, M., Lakshika, E.: Wearable sensors for recognizing individuals undertaking daily activities. In: Proceedings of the 2018 ACM International Symposium on Wearable Computers, pp. 64–67. ACM, New York (2018)
56. Sarela, A., Korhonen, I., Lotjonen, J., Sola, M., Myllymaki, M.: Ist vivago/spl reg/-an intelligent social and remote wellness monitoring system for the elderly. In: Information Technology Applications in Biomedicine, 2003. 4th International IEEE EMBS Special Topic Conference on, pp. 362–365. IEEE, Birmingham, UK (2003)
57. Sungmee, P., Sundaresan, J.: Enhancing the quality of life through wearable technology. IEEE Eng. Med. Biol. Mag. **22**(3), 41–48 (2003)
58. Hossain, T., Islam, M.S., Ahad, M.A.R., Inoue, S.: Human activity recognition using earable device. In: Proceedings of the 2019 ACM International Joint Conference on Pervasive and Ubiquitous Computing and Proceedings of the 2019 ACM International Symposium on Wearable Computers, pp. 81–84. ACM, New York (2019)
59. Kwon, H., Abowd, G.D., Plötz, T.: Adding structural characteristics to distribution-based accelerometer representations for activity recognition using wearables. In: Proceedings of the 2018 ACM International Symposium on Wearable Computers, pp. 72–75. ACM, New York (2018)
60. Davide, A., Alessandro, G., Luca, O., Parra, Xavier, F., Ortiz, L., Reyes, J.L.: Energy efficient smartphone-based activity recognition using fixed-point arithmetic. J. Univ. Comput. Sci. **19**(9), 1295–1314 (2013)
61. Xu, X., Tang, J., Zhang, X., Liu, X., Zhang, H., Qiu, Y.: Exploring techniques for vision based human activity recognition: methods, systems, and evaluation. Sensors **13**(2), 1635–1650 (2013)
62. Amirbandi, E.J., Shamsipour, G.: Exploring methods and systems for vision based human activity recognition. In: Swarm Intelligence and Evolutionary Computation (CSIEC), 2016 1st Conference on, pp. 160–164. IEEE (2016)

# Chapter 2
# Basic Structure for Human Activity Recognition Systems: Preprocessing and Segmentation

**Abstract** Automatic recognition of human activities using sensor-based systems is commonly known as human activity recognition (HAR). It is required to follow a structural pipeline to recognize activity using a machine learning technique. This chapter represents the different stages of this structural pipeline in detail. Following this, the preprocessing steps have been analyzed to clean and remove noises from raw sensor data. The importance of segmentation and criterions to select the best windowing method have been also described based on previous research works. The challenges regarding the selection of window length, window type, choosing overlapping percentage, and the relation between window duration and performance have been also investigated in the end.

## 2.1 Basic Structure for Human Activity Recognition Systems

Human activity is the summation of human action, where action can be defined as basic and the smallest element of something done by a human being [1, 2]. Human activity recognition is not a simple task as there can be a periodic occurrence of an action, two activities may have almost similar properties. Besides, it varies from person to person how an activity is performed. Moreover, occlusions, environmental noises, sensor orientation, and data acquisition issues make the task of accurate recognition more difficult [3–6]. There are several ways to derive information from raw sensor data about everyday human activities [7]. We can summarize the fundamental steps as pre-processing of raw sensor data, segmentation of filtered raw data, feature extraction and classification. Figure 2.1 depicts a generic architecture for sensor-based human activity recognition (HAR) systems.

© Springer Nature Switzerland AG 2021

M. A. R. Ahad et al., *IoT Sensor-Based Activity Recognition*, Intelligent Systems Reference Library 173, https://doi.org/10.1007/978-3-030-51379-5_2

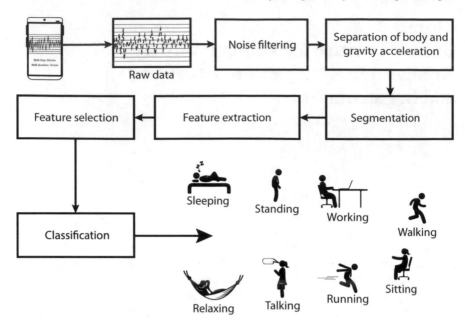

**Fig. 2.1** A simple structure for sensor-based human activity recognition

## 2.2 Pre-processing of Raw Sensor Data

### 2.2.1 Noise Filtering

The direct use of raw sensor data for further analysis is not a good idea, as the signal is made up of several components and there can be intrinsic noise components. The very first stage in pre-processing the data is noise filtering. Normally, four filters, such as a median filter, a low-pass filter, a discrete wavelet package shrinkage, and a Kalman filter as shown in Fig. 2.2 are used to eliminate acceleration and gyroscope noise [8–11]. However, when choosing one or more filters, we need to be careful about the signal-to-noise ratio (SNR), the correlation coefficient (R) between the filtered signal and the reference signal, the cut-off frequency, the waveform delay, the filter size, the window length, etc., as shown in Fig. 2.3.

Kalman filters normally show larger SNR and R-value. Following the order then comes median filter, discrete wavelet package shrinkage, and finally Butterworth low-pass filter (with waveform delay). The performance of a Butterworth low-pass filter can be improved by correcting the waveform delay by adjusting filter order and cut-off frequency. Real-time efficiency and waveform delay are two significant factors to be considered in the case of sensor-based activity recognition if we want to develop algorithms for filtering sensor noises. Because of the use of previous data only to estimate the current state, the Kalman filter shows good real-time perfor-

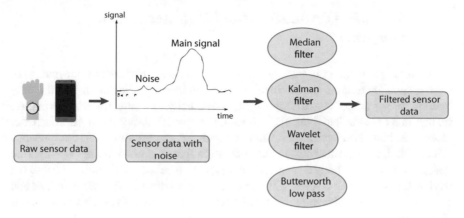

**Fig. 2.2** Filtering noises from raw sensor data

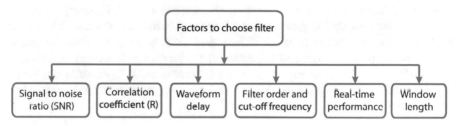

**Fig. 2.3** Factors to choose appropriate filter for sensor data

mance with a short delay. Performance of the median filter is related to its window length, N when applied in real-time. The reason behind this is the waiting time for about N/2 future data points to perform filtering. In case of discrete wavelet package shrinkage, decomposition level influences real-time performance. In spite of having little waveform delay, This filtering process requires to wait for at least $2^j$ for potential data points to eliminate the noise, where $j$ refers to decomposition level.

Butterworth filter is a casual system and does not need to wait for future data. But, we need to select the order of the filter and the cut-off frequency very carefully if we want to minimize waveform delay in the case of a Butterworth low-pass filter. By minimizing waveform delay for Butterworth filter, we can get better SNR and R-value than Kalman filter. Human activity frequency is usually about 0–20 Hz [12]. So, it is a good idea to choose the cut-off frequency of the Butterworth low pass filter to be 20 Hz to eliminate high-frequency noises. Filter order needs to be chosen based on the accuracy check, graphical analysis, and to ensure minimum waveform delay based on the sampling rate of data collection.

## 2.2.2 Separating Gravitational and Body Acceleration Component

Accelerometers respond to both gravitational and body acceleration but in case of a gyroscope, there is no gravitational part related to body orientation. So, the only reading we get from the gyroscope is due to the body movement. For more precise recognition of static and dynamic activities, if the accelerometer is used, it is better to separate body acceleration (BA) and gravitational acceleration (GA) components. We can get information about spatial orientation from the gravitational acceleration component whereas, body acceleration component is related to the movement of the device. So, we can use GA element related features to determine static activities and BA component related features to identify dynamic activities. These two components overlap in frequency.

A study in [8] has shown that while body acceleration has a frequency band from 0 to 20 Hz, most of the elements are above 0 Hz and below 3 Hz. This range overlaps the region covered by the gravitational component, which is generally between 0 and 1 Hz. This makes it difficult to differentiate those two components. So, based on accuracy check, Butterworth low-pass filter with corner frequency in between 0.1–0.5 Hz [13–15] can be used to separate GA component. Then subtracting this GA component from raw filtered data, we can get BA component. This process has been shown in Fig. 2.4.

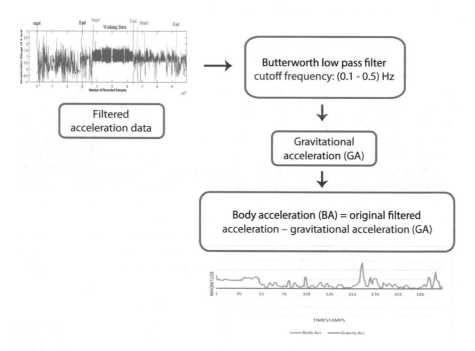

**Fig. 2.4** Separation of body acceleration (BA) and gravitational acceleration (GA)

### 2.2.3 Converting Categorical Data to Numerical Data

Categorical data are variables that contain values for the label rather than numerical values. Such categorical data do not help many of the learning algorithms. They require all numeric input and output variables. Therefore, we need to convert these categorical data into numerical data. There are two significant ways for this conversion namely:

- Integer encoding
- One-hot encoding

Integer encoding is the simplest encoding method and easily reversible. For each category, we can assign a unique integer value. For example, we can think about a dataset with three activity labels where the walk is 1, the run is 2, and the stand is 3. Integer encoding can serve the purpose for some variables due to the natural ordered relation between them. Most of the machine learning algorithms can recognize and link this relationship. But we can implement this encoding method only if there is an ordinal relationship exists between the variables. The ordinal relationship represents a sequence in which something is related to others of its kind. For example, there is a variation of features (mean and standard deviation of acceleration) for the mentioned activities which incorporates an ordinal relationship. To be more precise, the mean acceleration for the running activity will be higher than stand activity.

If there is no ordinal relationship between the variables like for categorical or nominal variables containing label values rather than numeric values, the integer encoding is not enough. If we still perform integer encoding for the variables without an ordinal relationship, the model will assume a natural ordering between categories may result in poor performance. In this case, we need to use the one-hot encoding. Instead of an integer variable, we can use binary variables to represent that data. One-hot encoding converts the categorical features into a format that fits best with algorithms for classification and regression. Some algorithms, such as random forests, manage native categorical values. Then, a hot encoding is not required.

## 2.3 Segmentation of Filtered Data

For continuous activity and motion detection, it's not easy for us to recover useful information from a continuous stream of sensor data. Therefore, we need to segment continuous stream into some segmented data form. To improve specific signal properties, various segmentation techniques can be used to obtain relevant information from a perpetual stream of data [16]. In this process, the sensor signal must be initially split into smaller time segments (windows). Later features are extracted from each window sample that is fed to the suitable classification algorithm. In the case of real-time data, windows need to be defined during the same period with data collection, which produces a perpetual real-time activity profile.

Segmentation is the process which divides sensor signals into smaller data segments. In the case of activity recognition field, most of the segmentation strategies can be classified into three groups:

1. Activity-defined windows
2. Event-defined windows
3. Sliding windows

### 2.3.1  Activity-Defined Windows

Activity-defined windowing procedure partitions sensor data stream based on the detection of changes in activity. Before identifying the specific activities explicitly, initial and endpoints need to be determined for each activity.

### 2.3.2  Event-Defined Windows

The basic technique of event-defined windowing approach is to locate specific events for further use. In this way, we can determine successive data partitioning. Preprocessing is required to locate specific events for event-defined windows. The events, in this case, may not be uniformly distributed in time. This is why window size does not play an important role. This analysis is mostly used in gait analysis along with the detection of heel strikes and toe-offs events. [17, 18] have presented some methods where windoing techniques are explored. Heel strike represents the initial floor contact while walking, and toe-offs represents the end of floor contact.

### 2.3.3  Sliding Windows

The sliding window approach which is referred to as "windowing", is the most extensively applied segmentation technique in the field of sensor-based human activity recognition. This method is simple for implementation and it requires less amount of preprocessing, that suits applications in real time. In this technique, the sensor signals are divided into windows of a fixed size and duration. This is your choice to keep no inter-window gaps or to choose a certain amount of overlap between adjacent windows. For some applications the overlap between the adjacent windows is allowed, where the activity is continuous enough and a sudden transition does not occur much.

Based on Banos et al. [19], the main contributions to each type of segmentation technique for on-body sensing activity recognition are summed up in Table 2.1.

**Table 2.1** Principal
segmentation techniques in
numerous research works

| Segmentation type | References |
|---|---|
| Activity-defined windows | [15, 20–29]. |
| Event-defined windows | [30–41]. |
| Sliding windows | [42–72]. |

### 2.3.3.1   Factors to Select Window Length

For the segmentation process, certain variables need to be considered when choosing window duration in sliding windows. A study [73] showed that the quality of extracted features are better from smaller windows as it creates the possibility to generate distinguishable features to separate activities. But due to more number of feature vectors from smaller size windows, it slowdowns the identification result rate for the end-user. Besides, if the activities are carried out for a short time, there is a high risk of identification failure in the event of longer windows. The optimal length of the window should be calculated depending on the operation being carried out. Moreover, we need to be careful enough while choosing the number of frames in case of windowing the sensor data. This is guided by a trade-off between two perspectives [74]:

- Information and
- Resolution.

When choosing the window length, long windows will include more information about an activity when a subject is performing only that single activity. This is how data is collected under laboratory set up. There is a common assumption behind the windowing technique while collecting data under lab set up that in general, a window will contain only a single activity no matter how long the window is. This hypothesis is likely to be breached in everyday life, where we can not control the activity changes and transitions.

### 2.3.3.2   Window Overlapping Issue

In previous studies, window size has been ranged from 0.08 second to 7 seconds mostly, which is shown in Table 2.2. Besides, some studies have indicated a fixed percentage of overlaps between neighboring windows [45, 75–77], whereas some studies preferred disjoint windows or non-overlapping windows [56, 78–80]. In the case of data points where each interval contains a piece of information that is unrelated to other intervals, non-overlapping windowing technique produces more precise results. Overlapping plays an important role when the signals at each interval are dependent of the other interval signals.

In case of processing the intervals separately, if the windows are not overlapping, we will be missing any information at the boundary of the windows. This is why we

**Table 2.2** Window length in numerous research works

| Window duration (second) | References |
| --- | --- |
| 0.08 | [81] |
| 1 | [78] |
| 1.5 | [79] |
| 3 | [80] |
| 5 | [56] |
| 7 | [82] |

use overlapping windows in short-time Fourier transforms (STFT) so that one of the windows will capture any jump in the signal curve and it will show in our analysis. If we use non-overlapping windows, we will miss the jumps in the signal curve and assume the signal is smooth. Studies have been showed that overlapping windows can handle the transitions more accurately [77]. On the contrary, to overcome the problem of misclassification due to transition, non-overlapping window duration should be small.

From [46], we have found that features derived from window lengths of one and two seconds on average achieve imperceptibly higher precision values than those with other window lengths. However, there are notable differences across different behaviors. If we can pick many lengths of windows for various tasks, this will lead to better rates of identification. From [46], we have also found that higher window times attain more precision for activities with higher velocity (e.g., the 2 s and 4 s window for skipping and hopping). For activities with moderate or lower velocities, windows with lower duration are preferred (e.g., the 1 s window for jogging and walking, and the 0.25 s and 0.5 s windows for standing and sitting).

## 2.4  Conclusion

In this chapter, we have summarized the basic architecture of a sensor-based human activity recognition system. The principal aspiration of this chapter is to show a detail explanation about the preprocessing steps of raw sensor data to get better performance. We have analyzed four types of noise filtering techniques with their pros and cons. Following this, the advantage of separating the input acceleration signal into the body and gravitational acceleration has been investigated. Finally, we have given a complete explanation of choosing the proper segmentation technique, their advantages, disadvantages, and parameter selection showing examples from previous research works.

## 2.5  Think Further

1. What is the general architecture of a Human Activity Recognition (HAR) system?
2. Should we explore raw data directly for classification? Why or why not?
3. What are the basic preprocessing steps of raw sensor data?
4. Which filters can be used to filter out noises from raw sensor data?
5. What are the factors that need to be taken care of while choosing filters?
6. What are the pros and cons of Kalman filter?
7. What are the pros and cons of median filter?
8. What are the pros and cons of wavelet package shrinkage filter?
9. What are the pros and cons of Butterworth low-pass filter?
10. How to choose the cut-off frequency of Butterworth low-pass filter?
11. What are the benefits of separating body acceleration (BA) and gravitational acceleration (GA)?
12. How to separate BA and GA component from acceleration data?
13. What are the importance of converting categorical data to numerical data?
14. In which cases we need to convert categorical data?
15. What are the importance of segmenting the filtered data?
16. What are the basic categories of segmentation process?
17. What are the basic requirements of choosing window length and window type?
18. What are the issues and benefits of using overlapping sliding window?
19. Based on the previous researches what should be the ideal window length to classify human activities?
20. What will be the differences in terms of performance in the case of a shorter window and longer window?

## References

1. Ahad, M.A.R.: Motion History Images for Action Recognition and Understanding. Springer Science & Business Media, Berlin (2012)
2. Ahad, M.A.R.: Computer Vision and Action Recognition: A Guide for Image Processing and Computer Vision Community for Action Understanding, vol. 5. Springer Science & Business Media, Berlin (2011)
3. Ahad, M.A.R.: Vision and sensor based human activity recognition: challenges ahead (2020)
4. Antar, A.D., Ahad, M.A.R., Shahid, O.: Vision-based action understanding for assistive healthcare: a short review. In: IEEE CVPR Workshop (2019)
5. Hossain, T., Goto, H., Ahad, M.A.R., Inoue, S.: A study on sensor-based activity recognition having missing data. In: 2018 Joint 7th International Conference on Informatics, Electronics & Vision (ICIEV) and 2018 2nd International Conference on Imaging, Vision & Pattern Recognition (icIVPR), pp. 556–561. IEEE, Kitakyushu (2018)
6. Antar, A.D., Ahmed, M., Ahad, M.A.R.: Challenges in sensor-based human activity recognition and a comparative analysis of benchmark datasets: a review. In: 2019 Joint 8th International Conference on Informatics, Electronics & Vision (ICIEV) and 2019 3rd International Conference on Imaging, Vision & Pattern Recognition (icIVPR), pp. 134–139. IEEE, Spokane, WA (2019)

7. Ghahramani, Z.: Unsupervised learning. In: Advanced Lectures on Machine Learning, pp. 72–112. Springer, Berlin (2004)
8. Mathie, M.: Monitoring and Interpreting Human Movement Patterns Using a Triaxial Accelerometer. University of New South Wales Sydney, Sydney (2003)
9. Huang, M., Zhao, G., Wang, L., Yang, F.: A pervasive simplified method for human movement pattern assessing. In: Parallel and Distributed Systems (ICPADS), 2010 IEEE 16th International Conference on, pp. 625–628. IEEE, Shanghai (2010)
10. Liu R., Zhou Ji., Liu M., Hou X.: A wearable acceleration sensor system for gait recognition. In: Industrial Electronics and Applications, 2007. ICIEA 2007. 2nd IEEE Conference on, pp. 2654–2659. IEEE, Harbin (2007)
11. Wen, T., Wang, L., Gu, J., Huang, B.: An acceleration-based control framework for interactive gaming. In: Engineering in Medicine and Biology Society, 2009. EMBC 2009. Annual International Conference of the IEEE, pp. 2388–2391. IEEE, Berlin (2009)
12. Antonsson, E.K., Mann, R.W.: The frequency content of gait. J. Biomechan. 18(1), 39–47 (1985)
13. Fahrenberg, J., Foerster, F., Smeja, M., Müller, W.: Assessment of posture and motion by multichannel piezoresistive accelerometer recordings. Psychophysiology 34(5), 607–612 (1997)
14. Foerster, F., Fahrenberg, J.: Motion pattern and posture: correctly assessed by calibrated accelerometers. Behav. Res. Methods Instrum. Comput. 32(3), 450–457 (2000)
15. Khan, A.M., Lee, Y.-K., Lee, S.Y., Kim, T.-S.: A triaxial accelerometer-based physical-activity recognition via augmented-signal features and a hierarchical recognizer. IEEE Trans. Inf. Technol. Biomed. 14(5), 1166–1172 (2010)
16. Avci, A., Bosch, S., Marin-Perianu, M., Marin-Perianu, R., Havinga, P.: Activity recognition using inertial sensing for healthcare, wellbeing and sports applications: a survey. In: Architecture of Computing Systems (ARCS), 2010 23rd International Conference on, pp. 1–10. VDE, Frankfurt (2010)
17. Antar, A.D., Ahmed, M., Hossain, T., Muramatsu, D., Makihara, Y., Inoue, S., Yagi, Y., Ahad, M.A.R., Ngo, T.T.: Wearable sensor-based gait analysis for age and gender estimation (2020)
18. Ngo, T.T., Ahad, A.R. Md., Antar, A.D., Ahmed, M., Muramatsu, D., Makihara, Y., Yagi, Y., Inoue, S., Hossain, T., Hattori, Y.: Ou-isir wearable sensor-based gait challenge: age and gender. In: Proceedings of the 12th IAPR International Conference on Biometrics, ICB (2019)
19. Banos, O., Galvez, J.-M., Damas, M., Pomares, H., Rojas, I.: Window size impact in human activity recognition. Sensors 14(4), 6474–6499 (2014)
20. Sekine, M., Tamura, T., Togawa, T., Fukui, Y.: Classification of waist-acceleration signals in a continuous walking record. Med. Eng. Phys. 22(4), 285–291 (2000)
21. Lester, J., Choudhury, T., Borriello, G.: A practical approach to recognizing physical activities. In: International Conference on Pervasive Computing, pp. 1–16. Springer, Berlin (2006)
22. Nyan, M.N., Tay, F.E.H., Seah, K.H.W., Sitoh, Y.Y.: Classification of gait patterns in the time-frequency domain. J. Biomech. 39(14), 2647–2656 (2006)
23. He, Z., Jin, L.: Activity recognition from acceleration data based on discrete consine transform and svm. In: Systems, Man and Cybernetics, 2009. SMC 2009. IEEE International Conference on, pp. 5041–5044. IEEE, San Antonio, TX (2009)
24. Gu, T., Wu, Z., Tao, X., Pung, H.K., Lu, J.: epsicar: an emerging patterns based approach to sequential, interleaved and concurrent activity recognition. In: Pervasive Computing and Communications, 2009. PerCom 2009. IEEE International Conference on, pp. 1–9. IEEE, Galveston, TX (2009)
25. Györbíró, N., Fábián, Á., Hományi, G.: An activity recognition system for mobile phones. Mob. Netw. Appl. 14(1), 82–91 (2009)
26. Hong, Y.-J., Kim, I.-J., Ahn, S.C., Kim, H.-G.: Mobile health monitoring system based on activity recognition using accelerometer. Simul. Model. Pract. Theory 18(4), 446–455 (2010)
27. Figo, D., Diniz, P.C., Ferreira, D.R., Cardoso, J.: Preprocessing techniques for context recognition from accelerometer data. Pers. Ubiquitous Comput. 14(7), 645–662 (2010)
28. Dernbach, S., Das, B., Krishnan, N.C., Thomas, B.L., Cook, D.J.: Simple and complex activity recognition through smart phones. In: Intelligent Environments (IE), 2012 8th International Conference on, pp. 214–221. IEEE, Roma (2012)

29. Yoshizawa, M., Takasaki, W., Ohmura, R.: Parameter exploration for response time reduction in accelerometer-based activity recognition. In: Proceedings of the 2013 ACM conference on Pervasive and ubiquitous computing adjunct publication, pp. 653–664. ACM, New York (2013)

30. Aminian, K., Rezakhanlou, K., De Andres, E., Fritsch, C., Leyvraz, P.-F., Robert, P.: Temporal feature estimation during walking using miniature accelerometers: an analysis of gait improvement after hip arthroplasty. Med. Biol. Eng. Comput. 37(6), 686–691 (1999)

31. Aminian, K., Najafi, B., Büla, C., Leyvraz, P.-F., Robert, Ph.: Spatio-temporal parameters of gait measured by an ambulatory system using miniature gyroscopes. J. Biomech. 35(5), 689–699 (2002)

32. Mansfield, A., Lyons, G.M.: The use of accelerometry to detect heel contact events for use as a sensor in fes assisted walking. Med. Eng. Phys. 25(10), 879–885 (2003)

33. Zijlstra, W., Hof, At.L.: Assessment of spatio-temporal gait parameters from trunk accelerations during human walking. Gait Posture 18(2), 1–10 (2003)

34. Zijlstra, W.: Assessment of spatio-temporal parameters during unconstrained walking. Eur. J. Appl. Physiol. 92(1–2), 39–44 (2004)

35. Selles, R.W., Formanoy, M.A.G., Bussmann, J.B.J., Janssens, P.J., Stam, H.J.: Automated estimation of initial and terminal contact timing using accelerometers; development and validation in transtibial amputees and controls. IEEE Trans. Neural Syst. Rehabil. Eng. 13(1), 81–88 (2005)

36. Jasiewicz, J.M., Allum, J.H.J., Middleton, J.W., Barriskill, A., Condie, P., Purcell, B., Li, R.C.T.: Gait event detection using linear accelerometers or angular velocity transducers in able-bodied and spinal-cord injured individuals. Gait Posture 24(4), 502–509 (2006)

37. Ward, J.A., Lukowicz, P., Troster, G., Starner, T.E.: Activity recognition of assembly tasks using body-worn microphones and accelerometers. IEEE Trans. Anal. Mach. Intell. 28(10), 1553–1567 (2006)

38. Benocci, M., Bächlin, M., Farella, E., Roggen, D., Benini, L., Tröster, G.: Wearable assistant for load monitoring: recognition of on—body load placement from gait alterations. In: Pervasive Computing Technologies for Healthcare (PervasiveHealth), 2010 4th International Conference on-NO PERMISSIONS, pp. 1–8. IEEE (2010)

39. Sant'Anna, A., Wickström, N.: A symbol-based approach to gait analysis from acceleration signals: identification and detection of gait events and a new measure of gait symmetry. IEEE Trans. Inf. Technol. Biomed. 14(5), 1180–1187 (2010)

40. Dobkin, B.H., Xu, X., Batalin, M., Thomas, S., Kaiser, W.: Reliability and validity of bilateral ankle accelerometer algorithms for activity recognition and walking speed after stroke. Stroke 42(8), 2246–2250 (2011)

41. Aung, M.S.H., Thies, S.B., Kenney, L.P.J., Howard, D., Selles, R.W., Findlow, A.H., Goulermas, J.Y.: Automated detection of instantaneous gait events using time frequency analysis and manifold embedding. IEEE Trans. Neural Syst. Rehabil. Eng. 21(6), 908–916 (2013)

42. Mantyjarvi, J., Himberg, J., Seppanen, T.: Recognizing human motion with multiple acceleration sensors. In: Systems, Man, and Cybernetics, 2001 IEEE International Conference on, vol. 2, pp. 747–752. IEEE, Tucson, AZ (2001)

43. Kern, N., Schiele, B., Schmidt, A.: Multi-sensor activity context detection for wearable computing. In: European Symposium on Ambient Intelligence, pp. 220–232. Springer, Berlin (2003)

44. Krause, A., Siewiorek, D.P., Smailagic, A., Farringdon, J.: Unsupervised, dynamic identification of physiological and activity context in wearable computing. In: null, p. 88. IEEE (2003)

45. Bao, L., Intille, S.S.: Activity recognition from user-annotated acceleration data. In: International Conference on Pervasive Computing, pp. 1–17. Springer, Berlin (2004)

46. Huynh, T., Schiele, B.: Analyzing features for activity recognition. In: Proceedings of the 2005 joint conference on Smart objects and ambient intelligence: innovative context-aware services: usages and technologies, pp. 159–163. ACM, New York (2005)

47. Maurer, U., Smailagic, A., Siewiorek, D.P., Deisher, M.: Activity recognition and monitoring using multiple sensors on different body positions. In: Wearable and Implantable Body Sensor Networks, 2006. BSN 2006. International Workshop on, p. 4. IEEE, Cambridge, MA (2006)

48. Parkka, J., Ermes, M., Korpipaa, P., Mantyjarvi, J., Peltola, J., Korhonen, I.: Activity classification using realistic data from wearable sensors. IEEE Trans. Inf. Technol. Biomed. **10**(1),
119–128 (2006)
49. Huynh, T., Blanke, U., Schiele, B.: Scalable recognition of daily activities with wearable sensors. In: International Symposium on Location-and Context-Awareness, pp. 50–67. Springer,
Berlin (2007)
50. Pirttikangas, S., Fujinami, K., Nakajima, T.: Feature selection and activity recognition from
wearable sensors. In: International Symposium on Ubiquitous Computing Systems, pp. 516–
527. Springer, Berlin (2006)
51. Wang, N., Ambikairajah, E., Lovell, N.H., Celler, B.G.: Accelerometry based classification
of walking patterns using time-frequency analysis. In: Engineering in Medicine and Biology
Society, 2007. EMBS 2007. 29th Annual International Conference of the IEEE, pp. 4899–4902.
IEEE, Arlington, VA (2007)
52. Suutala, J., Pirttikangas, S., Röning, J.: Discriminative temporal smoothing for activity recognition from wearable sensors. In: International Symposium on Ubiquitous Computing Systems,
pp. 182–195. Springer, Berlin (2007)
53. Amft, O., Tröster, G.: Recognition of dietary activity events using on-body sensors. Artif. Intel.
Med. **42**(2), 121–136 (2008)
54. Stikic, M., Huynh, T., Van Laerhoven, K., Schiele, B.: Adl recognition based on the combination of rfid and accelerometer sensing. In: Pervasive Computing Technologies for Healthcare,
2008. PervasiveHealth 2008. Second International Conference on, pp. 258–263. IEEE, Tampere
(2008)
55. Preece, S., Goulermas, J.Y., Kenney, L.P.J., Howard, D., et al.: A comparison of feature extraction methods for the classification of dynamic activities from accelerometer data. IEEE Trans.
Biomed. Eng. **56**, 871–879 (2009)
56. Altun, K., Barshan, B.: Human activity recognition using inertial/magnetic sensor units. In:
International Workshop on Human Behavior Understanding, pp. 38–51. Springer, Berlin (2010)
57. Han, C.W., Kang, S.J., Kim, N.S.: Implementation of hmm-based human activity recognition
using single triaxial accelerometer. IEICE Trans. Fundam. Electron. Commun. Comput. Sci.
**93**(7), 1379–1383 (2010)
58. Khan, A.M., Lee, Y.-K., Lee, S.Y., Kim, T.-S.: Human activity recognition via an accelerometer-
enabled-smartphone using kernel discriminant analysis. In: Future Information Technology
(FutureTech), 2010 5th International Conference on, pp. 1–6. IEEE, Busan (2010)
59. Marx, R.: Ad-hoc accelerometer activity recognition in the iball. In: Proceedings of the 2012
ACM Conference on Ubiquitous Computing, Pittsburgh, PA, USA, vol. 58 (2012)
60. Sun, L., Zhang, D., Li, B., Guo, B., Li, S.: Activity recognition on an accelerometer embedded mobile phone with varying positions and orientations. In: International Conference on
Ubiquitous Intelligence and Computing, pp. 548–562. Springer, Berlin (2010)
61. Atallah, L., Lo, B., King, R., Yang, G.-Z.: Sensor positioning for activity recognition using
wearable accelerometers. IEEE Trans. Biomed. Circuits Syst. **5**(4), 320–329 (2011)
62. Gjoreski, H., Gams, M.: Accelerometer data preparation for activity recognition. In: Proceedings of the International Multiconference Information Society, Ljubljana, Slovenia, vol. 1014,
p. 1014 (2011)
63. Jiang, M., Shang, H., Wang, Z., Li, H., Wang, Y.: A method to deal with installation errors of
wearable accelerometers for human activity recognition. Physiol. Meas. **32**(3), 347 (2011)
64. Kwapisz, J.R., Weiss, G.M., Moore, S.A.: Activity recognition using cell phone accelerometers.
ACM SigKDD Explor. Newsletter **12**(2), 74–82 (2011)
65. Lee, Y.-S., Cho, S.-B.: Activity recognition using hierarchical hidden markov models on a
smartphone with 3d accelerometer. In: International Conference on Hybrid Artificial Intelligence Systems, pp. 460–467. Springer, Berlin (2011)
66. Siirtola, P., Röning, J.: User-independent human activity recognition using a mobile phone:
offline recognition versus real-time on device recognition. In: Distributed Computing and Artificial Intelligence, pp. 617–627. Springer, Berlin (2012)

67. Wang, J.-H., Ding, J.-J., Chen, Y., Chen, H.-H.: Real time accelerometer-based gait recognition using adaptive windowed wavelet transforms. In: Circuits and Systems (APCCAS), 2012 IEEE Asia Pacific Conference on, pp. 591–594. IEEE, Kaohsiung (2012)
68. Hemalatha, C.S., Vaidehi, V.: Frequent bit pattern mining over tri-axial accelerometer data streams for recognizing human activities and detecting fall. Proc. Comput. Sci. **19**, 56–63 (2013)
69. Yunyoung Nam and Jung Wook Park: Physical activity recognition using a single triaxial accelerometer and a barometric sensor for baby and child care in a home environment. J. Ambient Intell. Smart Environ. **5**(4), 381–402 (2013)
70. Yunyoung Nam and Jung Wook Park: Child activity recognition based on cooperative fusion model of a triaxial accelerometer and a barometric pressure sensor. IEEE J. Biomed. Health Inform. **17**(2), 420–426 (2013)
71. Zheng, Y., Wong, W.-K., Guan, X., Trost, S.: Physical activity recognition from accelerometer data using a multi-scale ensemble method. In: IAAI (2013)
72. Mannini, A., Intille, S.S., Rosenberger, M., Sabatini, A.M., Haskell, W.: Activity recognition using a single accelerometer placed at the wrist or ankle. Med. Sci. Sports Exerc. **45**(11), 2193 (2013)
73. Khan, A.M.: Human activity recognition using a single tri axial accelerometer. Department of Computer Engineering, Graduate School, Kyung Hee University, Seoul, Korea (2011)
74. Schindler, K., Van Gool, L.: Action snippets: how many frames does human action recognition require? In: Computer Vision and Pattern Recognition, 2008. CVPR 2008. IEEE Conference on, pp. 1–8. IEEE, Anchorage (2008)
75. Preece, S.J., Goulermas, J.Y., Kenney, L.P.J., Howard, D., Meijer, K., Crompton, R.: Activity identification using body-mounted sensors—a review of classification techniques. Physiol. Meas. **30**(4), R1 (2009)
76. Tapia, E.M., Intille, S.S., Haskell, W., Larson, K., Wright, J., King, A., Friedman, R.: Real-time recognition of physical activities and their intensities using wireless accelerometers and a heart rate monitor. In: Wearable Computers, 2007 11th IEEE International Symposium on, pp. 37–40. IEEE, Washington, D.C. (2007)
77. Lara, O.D., Pérez, A.J., Labrador, M.A., Posada, J.D.: Centinela: a human activity recognition system based on acceleration and vital sign data. Pervas. Mobile Comput. **8**(5), 717–729 (2012)
78. Reddy, S., Mun, M., Burke, J., Estrin, D., Hansen, M., Srivastava, M.: Using mobile phones to determine transportation modes. ACM Trans. Sensor Netw. (TOSN) **6**(2), 13 (2010)
79. Cheng, J., Amft, O., Lukowicz, P.: Active capacitive sensing: exploring a new wearable sensing modality for activity recognition. In: International Conference on Pervasive Computing, pp. 319–336. Springer, Berlin (2010)
80. Minnen, D., Westeyn, T., Ashbrook, D., Presti, P., Starner, T.: Recognizing soldier activities in the field. In: 4th International Workshop on Wearable and Implantable Body Sensor Networks (BSN 2007), pp. 236–241. Springer, Berlin (2007)
81. Berchtold, M., Budde, M., Schmidtke, H.R., Beigl, M.: An extensible modular recognition concept that makes activity recognition practical. In: Annual Conference on Artificial Intelligence, pp. 400–409. Springer, Berlin (2010)
82. McGlynn, D., Madden, M.G.: An ensemble dynamic time warping classifier with application to activity recognition. In: Research and Development in Intelligent Systems xxvii, pp. 339–352. Springer, Berlin (2011)

# Chapter 3
# Methodology of Activity Recognition: Features and Learning Methods

**Abstract** Sensor-based Human Activity Recognition (HAR) has been explored by many research communities and industries for various applications. Conventional pattern recognition approaches based on handcrafted features contributed a lot in this research field by employing general classification approaches. This chapter represents those handcrafted features in time and frequency domain along with their importance and feature selection methods. Following these methods, this chapter provides explanations of several conventional machine learning techniques for classifying sensor data for activity recognition. The problems of overfitting and underfitting have been discussed with remedies. Previous research works using conventional pattern recognition (PR) approaches on some benchmark datasets have also been analyzed.

## 3.1 Introduction on Methods of Activity Recognition

Sensor-based Human Activity Recognition (HAR) has been explored by many research communities and industries for various applications [1–5]. After some preprocessing steps and segmentation (if it is required to segment temporally, based on the available dataset) of raw sensor data, the feature extraction is done from the processed data. Then, we need to create an algorithm pipeline and model by using machine learning techniques from the set of feature instances. Moreover, it is required to train the model that will recognize the activities using the sensor data.

## 3.2 Features Extraction

In the field of Human Activity Recognition (HAR), due to the oscillatory and highly fluctuating nature of raw accelerometer signals, as shown in Fig. 3.1, correct identification of activity patterns gets complicated. In real life, people normally perform similar movements in numerous ways. This can create a large variation in the extracted features in real-time. To overcome this problem, after the segmentation stage, we can concentrate to extract features that are more relevant and robust, from the seg-

© Springer Nature Switzerland AG 2021                                                                                     27
M. A. R. Ahad et al., *IoT Sensor-Based Activity Recognition*, Intelligent Systems
Reference Library 173, https://doi.org/10.1007/978-3-030-51379-5_3

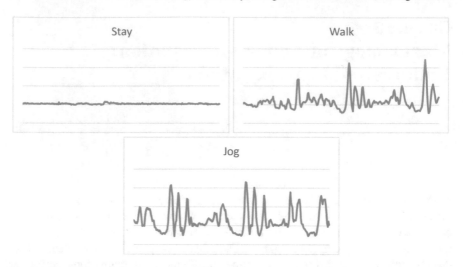

**Fig. 3.1** Acceleration signals for various activities

mented data. These can be useful in differentiating the activities that are correlated to each other. In the case of large-scale data, this is necessary to find a high-level representation so that we can ensure the generalization capacity of the HAR system.

We can regard a feature set as properly extracted feature vector if the feature set shows a small contrast between replications of the identical actions and across various subjects. Also, the feature set should alter considerably between various activities.

In most of the research works, existing human activity recognition systems based on accelerometer data use the statistical approach of feature extraction. Most of the common features used for activity recognition are the mean and standard deviation. But there are lots of other features. We can classify these features into the following categories in general:

1. Time domain or statistical domain features
2. Frequency domain or spectral domain features

Figure 3.2 presents the list of features analyzed in previous works. We have shown a basic block diagram of feature extraction technique in Fig. 3.3.

## 3.2.1 Time Domain Features

In most of the statistical analysis, an elementary task is to distinguish the location and variability of the time series data. Time-domain features are usually extracted directly from a patch of time series data called 'window'. Different types of time domain features have been extensively explored, e.g., in [6–8], [9–15]. In [16], *mean* and *standard deviation* have been explored for activity recognition. We can represent

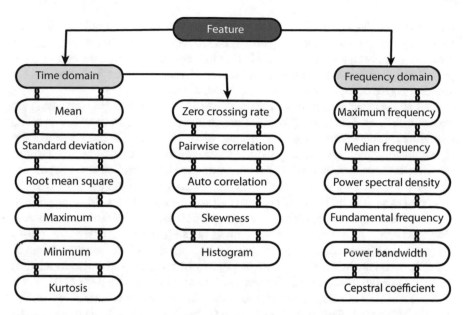

**Fig. 3.2** Time domain and frequency domain features for sensor-based activity recognition

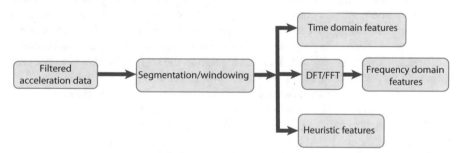

**Fig. 3.3** A basic block diagram of feature extraction technique

*mean* value as the DC component or average value of the signal over a particular window. The stability of a time series signal can be represented by the *standard deviation*, which measures the variability of the signal over a window. We can use the standard deviation value of the acceleration signal (from accelerometer) to capture the range of possible acceleration values. This can help to distinguish activities or behaviors that may appear identical in appearance but vary in pace and strength (e.g., walking versus jogging versus running). The *mean absolute deviation* (MAD) was traditionally used to automatically diagnose epileptic seizures [6].

For the differentiation of static activities (e.g., sitting, lying, etc.) and dynamic activities (e.g., walk, jog, etc.) and identification of postures, *mean* plays an important role [7]. There are other features named *variance, median, skewness, interquartile range, kurtosis, root mean square* (RMS), etc. These features have been widely used by many researchers [9, 13, 17, 18]. Kurtosis feature denotes the peak of a curve of a frequency distribution. When using this feature, the estimation of whether the data is elevated or flat compared to a normal distribution is possible. In fact, the tailedness of probability distribution of a random variable (real-valued) can be determined along with the degree of asymmetry of the sensor signal distribution [19].

The degree of asymmetry of the sensor data distribution can be measured using the *skewness* feature. This function's inputs are an n-dimensional array with data along with the axis, along which skewness is computed. The quadratic mean value of the signal is called *root mean square* (RMS) value, which is calculated over a window. In previous researches, this feature has been exploited to distinguish walking patterns [20]. RMS value feature is popular too in current activity recognition works [21].

A measure of the statistical dispersion is represented by interquartile range (IQR). The difference between the 75th and the 25th percentiles of a signal over a window is similar to IQR. The interquartile range depicts the dispersal of the data and bypasses the impact on range produced by extreme values in the data when the mean values of various classes are alike. This feature has shown its usefulness in epileptic seizures detection and fall detection [6], likewise median absolute deviation (MAD).

If we want to estimate the intensity and direction of a linear relationship between two different signals then, *signal correlation* feature is one of the best choices. In the case of identifying activities, this feature distinguishes between activities that include translation in a particular dimension [22]. To determine the degree of correlation, it is important to measure the coefficients of correlation between the signals along different axes. *Pearson's coefficient*, which is a specimen correlation coefficient is profoundly used in this manner. It can computed as the ratio of the covariance of the signals along the x-axis and the y-axis to the product of their standard deviations [23]. In addition, the *cross-correlation coefficient* has been used to measure the relationship between acceleration signals from different axes on the same body segment and across different segments [24]. To estimate the self-similarity of time series segments, we can use the auto-correlation feature in case of activity recognition [25].

*Zero-crossing* is another temporal domain feature. We can define this feature as a normalized (by the window length) value of the total number of times that the signal changes from positive to negative, or the reverse. This feature has been applied profoundly in both speech recognition and music information retrieval. This is a key feature to recognize the neighboring circumstances or the type of sounds such as music, speech, and noise [26]. Zero crossing rate can also be used in human activity recognition perspective [27].

There are some other features used in different work such as *integral of modulus of accelerations* [13], *jerk* [28, 29]. Jerk is the derivative of acceleration, which measures the rate of change of acceleration. When a phone is carried in pocket or handbag, the orientation of the accelerometer is generally unknown. In these cases, it is difficult to isolate body-related accelerations from the gravitational acceleration, or discover the real directions of the perceived accelerations correctly. As an explication, the jerk feature (i.e., changes of accelerations) is explored rather than the primary acceleration signal. The total jerk magnitude is fully independent of any sensors' orientations. However, it can represent the body-related accelerations. The

**Table 3.1**  Different types of features and their derivation formulas

| Feature | Equation |
|---|---|
| Mean | $\dfrac{\left(\sum_{i=0}^{n-1} A_i\right)}{n}$ |
| Standard deviation | $\dfrac{\sum_{i=0}^{n-1}|A_{mean}-A_i|}{\sqrt{n}}$ |
| Variance | $\dfrac{\sum_{i=0}^{n-1}(A_{mean}-A_i)^2}{n}$ |
| Skewness | $\dfrac{n}{(n-1)(n-2)}\sum_{i=0}^{n-1}\left(\dfrac{A_i-A_{mean}}{A_{std}}\right)^3$ |
| Kurtosis | $\left[\dfrac{n(n+1)}{(n-1)(n-2)(n-3)}\sum_{i=0}^{n-1}\left(\dfrac{A_i-A_{mean}}{A_{std}}\right)^4\right]-\dfrac{3(n-1)^2}{(n-2)(n-3)}$ |
| Inter-quartile range | $A_{mod\ of\ last\ half} - A_{mod\ of\ first\ half}$ |
| Average peak- trough- distance | $\dfrac{\sum |A_{relative\ maxima}-A_{relative\ minima}(adjacent)|}{m}$ |
| Integral of modulus of accelerations | $\int_0^{n-1}|a_x|dt + \int_0^{n-1}|a_y|dt + \int_0^{n-1}|a_z|dt$ |
| Parameters | $n$ = window size, $A_i$ = amplitude of (ith) point of the window, $std$ = standard deviation, $m$ = total number of maxima to minima or minima to maxima path. |

jerk signal can be improved further with specific knowledge on jerk angles (directional changes) if the direction of gravitation can be approximated [30]. In Table 3.1, we have summarized some of the features and their derivation formula.

## 3.2.2   Frequency Domain Features

To extract frequency-domain features, the *Fast Fourier Transform* (FFT) is a broadly used transformation method to reconstruct window sensor data in the frequency domain. A set of coefficients are found from the output of an FFT, which denotes the amplitudes of the signal frequency components and the distribution of signal energy. We can explore median frequency [31] and subsets of the various FFT coefficients [32] to derive the spectral distribution from those coefficients.

Some other features can be extracted from this FFT series. One of this feature is *spectral energy*. It can be computed from the summations of the squared FFT coefficients [33]. If we sum up the normalized information entropy of the FFT components, that can be another feature named entropy [34]. Entropy differentiates activities with simple and complex acceleration patterns. A good example in this regard can be the cycling activity, which requires uniform movement of legs. If we perform frequency-domain analysis of thigh acceleration, it displays a single dominant frequency. In opposite, running activity is composed of more complex acceleration pattern and it often shows many major FFT components. These variations lead to a much higher frequency domain entropy for running than cycling if we compare [33]. To differentiate activities with single dominant frequency and activities with complex acceleration pattern with many dominant FFT components, frequency domain analysis is a good option.

There are some other methods of feature extraction which are not very much common but used in additional cases to improve recognition accuracy. These types of features are described below.

## 3.3   Wavelet Analysis-Based Features

Although Fourier analysis is usually used to collect information about a signal's frequency content, wavelet analysis can be employed to analyze both time and frequency characteristics. We can also examine fast-changing transient signals using wavelet analysis. Wavelets can be used to provide more precisely localized temporal and frequency information. By using this feature and utilizing a Fourier transform, we can formulate many applications. The formulation of wavelet analysis is done using continuous or discrete wavelet transform. The discrete wavelet transform (DWT) is typically computing by using a filter bank, where the initial signal is successively disintegrated into independent low- and high- pass filtered signals that can be considered approximations and coefficients.

We can get several unique coefficients, each of which includes data on a particular frequency band by decomposing a body-worn or smartphone sensor signal using wavelet analysis. These coefficients comprise information on temporal changes in frequency content. The reason behind this is the characterization of the original signal along its entire length by these coefficients. Therefore, a wavelet technique can be explored for evaluating and characterizing non-stationary signals with changing frequency context over time, which is not possible using Fourier analysis. In the case of activity monitoring, wavelet analysis has been mostly implemented in three constraints: identification of activity transition points [35], signal enhancement [36], and generation of time-frequency features consequently used for classification [20, 35].

## 3.4   Heuristic Features

We can represent the human movement or action patterns as time-varying segmental accelerations. Previous research works implemented various methods to derive certain heuristic features. The goal was to quantify the amplitude of these accelerations.

It is required to remove the baseline offset before deriving these features. We can do that by using a high pass filter. These features include,

- Signal magnitude area (SMA) [37],
- Peak-to-peak acceleration [38],
- Mean rectified value [39], and
- Root mean square [40].

These features are generally employed to distinguish static activities (e.g., sit, sleep, stand, etc.) from dynamic activities (e.g., walk, run, jog, etc.) [37]. When a subject is at rest (static acceleration due to gravity), the computed acceleration is equal to the cosine of the sensor orientation angle relative to the vertical. This angle can be mentioned as the *tilt angle*. The tilt angle can be used as an input to a classification solution, especially to the case where static postures are to be separated [41], and to recognize postural transitions [36].


## 3.5  Feature Selection and Dimensionality Reduction Methods

Generally, we act several identical actions in a variety of ways. This can generate a large variability in the features, which are extracted from on-body or smartphone sensor data. Features are varied too and different features can extract different kinds of information—hence, aid to classify activities. Besides, choice of less important features with redundant or irrelevant information can hamper recognition accuracy. So, it is important to find the importance and robustness of features before using them to build the learning model. If we need to deal with high dimensional space of features, dimensionality reduction is the most popular step in machine learning, which is done either by creating new dimensions or we can also select a subset of the original dimension. In this case, we can map the original feature space onto a new feature space which has a lower dimension [42].

We can pick the features depending on the element of value, which can be identified using Random Forest attribute. Another popular method is *Maximum Relevance and Minimum Redundancy* (MRMR) [43]. This method had been used in [44], where the minimum mutual information between features had been used as a criterion for minimum redundancy. Besides, for maximum relevance, the maximal mutual information between the classes and features had been used. Moreover, *Correlation-based Feature Selection* (CFS) method [45] had been also used in [46]. The CFS method works based on a primary assumption that features should be strongly correlated with the specified class, however, they must be uncorrelated with each other. Another feature selection method is a *forward-backward search*, in which features are added sequentially, but deleted from a larger set. It is also possible to identify the optimal features based on the produced classification results for each subset of the features. We summarize some feature selection techniques in this section. Figure 3.4 summarizes the concepts.

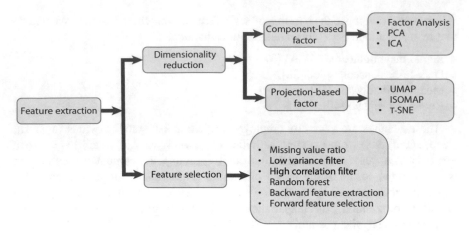

**Fig. 3.4**   Feature selection and dimensionality reduction techniques

### 3.5.1   Removing Feature with Low Variance

In this method, a variance threshold is set for features. Features whose variance does not meet the threshold are removed. In the default case, features with zero variance are removed that have the same value in all samples [47].

### 3.5.2   Missing Value Ratio

If a dataset contains more than 50% of missing values, we can either impute the missing values or we can drop the variables. Presence of missing value ensures that we will not have much information. This is the reason for dropping those values. It is required to set a threshold value first in this technique. When the amount of missing values in any variable surpasses that threshold value, we can remove the variable [47].

### 3.5.3   Univariate Feature Selection

This method selects the best features on the basis of the univariate statistical tests, e.g., false positive rate, false discovery rate, family-wise error, etc. It is regarded as a preprocessing step before estimating the result by the model. This method has some subcategories [48].

### 3.5.4 Select-k-Best

This method keeps k-Highest scoring features and eliminates the rest. The value of the k needs to be selected based on the characteristics of the feature sets [49].

### 3.5.5 Select Percentile

This method keeps a user-specified specific percentage of features and removes the rest [50].

### 3.5.6 Generic Univariate Selection

In this method, univariate feature selection is performed with a configurable strategy. This permits to choose the suitable univariate selection approach with hyperparameter search estimator [48].

### 3.5.7 Recursive Feature Elimination

This method works by recursively eliminating some attributes and building model based on the rest of the attributes. It utilizes the model efficiency to recognize which attributes or combination of attributes offer the most to predicting the target attribute. Firstly, the estimator is trained using the initial feature set and the importance of each feature is obtained. Then the less important features are eliminated from the feature set and this method is recursively iterated on the pruned set until the aspired number of features to select is ultimately attained [51].

### 3.5.8 L1-Based Feature Selection

Linear models castigated with the L1 norm have sparse solutions. In most of the cases, many of their estimated coefficients are zero. Based on this method, we can design an algorithm to select only the non-zero coefficients, which will reduce the dimensionality of the data [52].

### 3.5.9   T-Distributed Stochastic Neighbor Embedding (t-SNE)

This is an advanced and non-linear way to search for patterns. It is possible at the same time to recall both local and global structures of the data using the t-SNE algorithm. Besides, in both high and low dimensional space, this algorithm computes the probability similarity of points. The distinction between these two probabilities is minimized afterwards. Though in the case of big datasets this algorithm functions very well, it has some limitations. These limitations include delayed computation time and loss of large-scale information. This algorithm is also unable to represent huge datasets with lots of variables [53].

### 3.5.10   Uniform Manifold Approximation and Projection (UMAP)

As we mentioned earlier, the t-SNE algorithm has some limitations in the case of large datasets. Uniform Manifold Approximation and Projection (UMAP) algorithm, on the other hand, performs well as a dimensionality reduction technique than t-SNE. This algorithm can preserve both local and global structure in a higher amount than t-SNE. The runtime is also more abbreviated for UMAP. When we need to deal with multivariate and high dimensional large datasets, UMAP is preferred. The combination of visualization power and reduction of dimensions of the data has made UMAP a powerful dimensionality reduction technique for preserving both local and global structure of the data. It is based on Riemannian geometry as well as algebraic topology. This dimensional technique can map nearby points on the manifold to the points in the lower-dimensional presentation. It can perform similarly for the faraway points. This UMAP is based on an approximation of the k-nearest neighbor computation. An efficient optimization are accomplished by using the stochastic gradient descent algorithm [54]. For visualization quality, the UMAP has demonstrated comparable performances with the t-SNE. Moreover, it is free from any computational constraints on embedding dimensions [54].

### 3.5.11   Factor Analysis

Factor analysis technique groups the variable by using their correlations. In this technique, a particular group will contain only those variables having a higher amount of correlation among themselves, and a lower amount of correlation with other group variables. In this technique, each group is named as a *factor*. Though these factors or groups are lower in number than the original data dimensions, these factors are difficult to observe [55].

## 3.5.12 ISOMAP

ISOMAP (Isometric Feature Mapping) algorithm is a non-linear feature reduction approach. ISOMAP focuses on to obtain a low-dimensional representation from a non-linear manifold. This algorithm is different than the Principle Component Analysis (PCA) approach. The primary consideration of this technique is to assume the manifold smooth. Besides, it also implies that the geodesic distance between the two points is equal to that of the Euclidean distance for every two points on the manifold. Geodesic distance is known as the shortest or nearest distance between two points on a curved surface. On the other hand, Euclidean distance is the shortest path between two points on a straight line [56]. ISOMAP can be explored in a situation when higher-dimensional data and lower-dimensional manifold have a non-linear mapping.

## 3.5.13 Elimination of Highly Correlated Features

We can calculate the correlation coefficient among the features and set a correlation coefficient threshold. If any two feature has a correlation coefficient that crosses the threshold, we can consider them as the same feature and we can remove one [57].

## 3.5.14 Tree-Based Feature Selection

Tree-based estimators can be applied to calculate feature importance and to drop irrelevant features. The optimal condition is chosen based on *Gini Impurity*. Gini Impurity measures how frequently a randomly selected element from the set will be erroneously classified or labeled—if it were randomly labeled as per the order of labels in the subset. When we need to train a tree (or decision tree), it is required to measure how much each feature reduces the weighted impurity in a tree. Then the impurity reduction from each feature is averaged for forest and the features are ranked according to this measure. For forest, the impurity reduction from each feature is averaged and the features are ranked according to this measure [58].

In some other cases, we can consolidate the primary features to describe a new set of variables as an alternative, and choose a subset of the prevailing features. Applying this process can give us benefit in two ways:

- We can reduce the disproportionately produced features from many sensors.
- The newly decreased set of variables generally demonstrates more reliable discriminative ability while classification.

Two most widely utilized dimensionality reduction techniques in the arena of human activity recognition or monitoring using accelerometers are: PCA (Principle Component Analysis) and ICA (Independent Component Analysis) [59]. PCA procedure utilizes an orthogonal linear transformation that turns the data to a new set of observations or of likely correlated variables, to compute *principal components*. This method can emphasize variation and extract strong data set or patterns from the original data. ICA is a statistical and computational method. This procedure reveals concealed factors that carry sets of random variables, signals, or measurements. Moreover, *Discrete Cosine Transform* (DCT) has been also adopted in some works [60] along with autoregressive model coefficients [61], which obtained good recognition accuracies.


## 3.6    Choosing Appropriate Feature Selection or Dimensionality Reduction Technique

To remove superfluous and irrelevant data, increasing the comprehensibility of the result, and increasing learning accuracy dimensionality reduction plays an important role in the field of machine learning [62]. Classification accuracy is also closely related to several attribute space reduction techniques, which shows the significance of appropriate feature selection strategy [63]. In this section, we have shortly summarized the use cases of dimensionality reduction and we have tried to give an idea about which dimensionality reduction or feature selection techniques are good in terms of processing time and computational cost.

- **Missing value ratio**: To reduce the number of variables for a dataset containing too many missing values, this method can be used. If we find a large number of missing values in data, we can remove those variables.
- **Low variance filter method**: To separate and release constant variables from the dataset, we can employ the low variance filter method. The target variable is less influenced by the variables with lower variance. Therefore, we can reliably remove these variables.
- **High correlation filter method**: The multicollinearity is increased in a dataset if it contains high correlation variable pairs. High correlation filter method can identify the features that are highly correlated, and then remove them.
- **Random Forest (RnF) method**: The importance of the extracted feature can be found using the Random Forest technique, which is used by many researchers. It is possible to keep the topmost feature based on the feature importance rank, which reduces the dimensionality of the dataset.
- **Factor analysis technique**: For highly correlated sets of variables, factor analysis is the most preferred technique of dimensionality reduction. By utilizing this method, it is possible to group the variables based on the correlation of variables in a group and correlation of variables among different groups. Finally, each group is represented as a factor.
- **Principal Component Analysis (PCA) technique**: PCA is mostly used to deal with linear data. In this method, the data is divided into a set of components. The goal is to describe as much variance as possible.

- **Backward Feature Elimination and Forward Feature Selection method**: Because of higher computational time, both of these techniques are normally used on smaller datasets with a lower amount of variables.
- **Uniform Manifold Approximation and Projection (UMAP) algorithm**: UMAP algorithm shows better performance for high dimensional data. This method has an advantage over t-SNE due to shorter runtime.
- **ISOMAP algorithm**: ISOMAP algorithm performs well for strongly non-linear data by recovering full low-dimensional representation.
- **t-Distributed Stochastic Neighbor Embedding (t-SNE) algorithm**: The t-SNE algorithm also serves properly when the data is strongly non-linear. The performance is remarkably well for visualizations as well.
- **Independent Component Analysis (ICA) method**: ICA transforms the data into independent components. These independent components explain the data using a smaller quantity of components.

## 3.7 Feature Normalization

Extracted features always vary in magnitudes and it can create difficulties for some machine learning methodologies, where feature with greater magnitude has higher importance. This is not justified in all cases. To overcome this untoward effect, a normalizing step is necessary before classification. Because of the differences in scale factors and units among different features, all features should be normalized before proceeding to the feature selection stage. There are generally four types of feature scaling or normalizing methods:

1. **Rescaling (min-max normalization)**: This is the simplest method of feature scaling. This method rescales the range of features to scale the range in $[0, 1]$ or $[-1, 1]$. Based on the characteristics of the data, we can select the target range. The equation of the rescaling method can be shown as:

$$x_{norm} = \frac{x - min(x)}{max(x) - min(x)} \qquad (3.1)$$

where, $x =$ original value and $x_{norm} =$ normalized value.

2. **Mean normalization**: This is another normalization method where the average of the feature vector is subtracted from the original feature vector. Then the result is divided by the range to get the normalized feature value. The equation of the mean normalization method can be shown as:

$$x_{norm} = \frac{x - mean(x)}{max(x) - min(x)} \qquad (3.2)$$

where, $x$ = original value and $x_{norm}$ = normalized value.

3. **Standardization**: This is the most extensively used normalizing method in machine learning where we need to handle various data that can include multiple dimensions. In different machine learning methods like Support Vector Machine (SVM), Artificial Neural Network (ANN), Loristic Regression, etc.— this standardization approach is extensively employed for normalization [64]. In this process, when we subtract the mean in the numerator, each feature value in the data is required to have zero-mean and unit variance. The common practice of this technique is to compute the distribution mean as well as the standard deviation for each feature. After this step, the mean is subtracted from the corresponding feature vector. Finally, the result is divided by the standard deviation to get the normalized output. The formula is given below:

$$x_{norm} = \frac{x - mean(x)}{\sigma} \tag{3.3}$$

where, $x$ = original value, $x_{norm}$ = normalized value, and $\sigma$ = standard deviation.

4. **Scaling to unit length**: This is another common method of feature scaling. In this method, it is necessary to scale the components of a feature vector to achieve the outcome that the complete vector has length one. This normally indicates that we need to divide each component by the Euclidean length of the vector. The formula is:

$$x_{norm} = \frac{x}{||x||} \tag{3.4}$$

where, $x$ = original value, $x_{norm}$ = normalized value, and $||x||$ = Euclidean length of the vector.

In some application areas (e.g., Histogram features), using the L1 norm (i.e., snake distance, Manhattan Length, or Manhattan Distance, City-Block Length) of the feature vector is an effective strategy. This technique becomes more prominent and useful if a Scalar Metric is used as a distance measure in the following learning steps. We have summarized the feature normalization techniques in Fig. 3.5.

## 3.8  Learning

We can summarize the four different types of machine learning algorithms for labeled and unlabeled data. These primary techniques are:

- Supervised learning

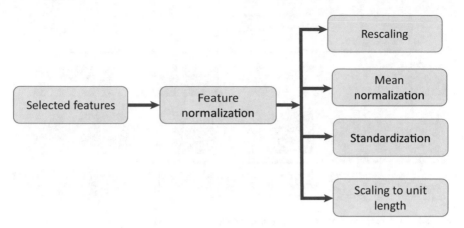

**Fig. 3.5** Feature normalization techniques

- Unsupervised learning
- Semi-supervised learning
- Reinforcement learning

Supervised learning is used for labeled data, unsupervised learning is used for unlabeled data, and semi-supervised learning is used for partially unlabeled data. In the case of reinforcement learning, algorithms learn to react to an environment. We have summarized the machine learning techniques in Fig. 3.6. In the case of human activity recognition, most of the sensor data are labeled, which is the primary reason for using a supervised learning technique. We can utilize various classification methods based on the amount of data, the type of data, similarities of activity classes, amount of activities, number of classes, etc. Some of the most common and widely-explored classifiers to classify dynmic and statis activities are:

- Linear Discriminant Analysis (LDA): The tolerance value for LDA has significant importance for the performance.
- Support Vector Machine (SVM): The selection of kernels for SVM is important.
- Logistic Regression (LR),
- K-Nearest Neighbour (KNN): The determination of K-value and the distance computation strategy for KNN are significant.
- Random Forest (RnF): Proper selection of the number of tress is needed.

The field of smartphone sensor-based activity recognition using a supervised machine learning technique includes many research works. In the case of classification technique, labeled data (most of the time labeled manually) are used. There is a high chance of generating training dataset with inappropriate labeling, which deteriorates the performance in real-time applications. To mitigate this problem, a semi-supervised learning-based recognition method is proposed that is dependent on self-training procedure [65]. However, they have employed their method on a small

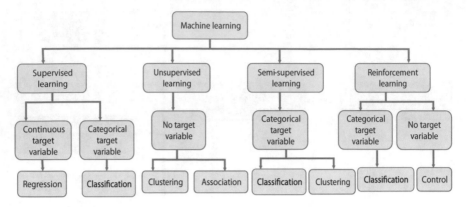

**Fig. 3.6**  Different types of machine learning techniques

sample of labeled training data. In another unsupervised approach based on K-Means clustering, activity recognition was accomplished in [66]. But this method failed to perfectly distinguish between sitting and standing activities (failure rate is higher: 37%). In the case of [67], they used three separate networks for three separate states (static activities, dynamic activities, and transitional activities) where the output of Linear Discriminant Analysis module was used as input to these networks. They achieved 85% accuracy for their owned dataset.

## 3.9  Machine Learning Techniques Related to Activity Recognition

Machine learning algorithms have become the most extensively used approaches in the activity recognition process based on the feature representation of data from an accelerometer alone or both accelerometer and gyroscope [68]. Various assumptions are made about the structure and shape of the function by various machine learning methods. We need to emphasis on streamlining a representation to approximate it. As mentioned earlier, there are four categories of machine learning algorithms. Among them, in most of the research works, supervised classification method has been utilized as most of the activity data are labeled and the task is to classify the activities. Besides, the output variables (activities) are mostly categorical and includes time series prediction, which is the core reason for choosing the classification (supervised learning) method. Earlier research works on sensor data focused on using one of the following approaches for training:

- Linear (assumed functional form is a linear combination of the input variables),
- Parametric (mapping is done to a known functional form),
- Nonlinear or non-parametric (able to learn any mapping from input to output), or
- Ensemble algorithms.

Short descriptions of the most common algorithms have been given in the next sub-sections.

### 3.9.1 Nonlinear or Non-parametric Algorithms

These algorithms do not constitute powerful hypotheses about the nature of the mapping function. Besides, these algorithms are also free to learn any functional form from the training data. In the case of non-parametric methods, there are no much worries about choosing just the right features. These algorithms can fit a large number of functional forms. This results in greater performance models for prediction and makes no assumptions or weak assumptions about the underlying function.

In spite of advantages, these methods have some primary limitations, namely—(i) they require a lot more training data for the mapping function estimation; (ii) they are a lot slower to train due to a large number of parameters to train; and (iii) there is a risk of overfitting the training data [69]. We have summarized the non-parametric algorithms that can be used for HAR in Fig. 3.7.

#### 3.9.1.1 Decision Tree (DT)

For predictive modeling machine learning, Decision Tree is an essential type of algorithm. However, in the case of decision trees, a non-linear relationship between predictors and outcome deteriorate the accuracy. Classification and Regression Trees (CART model) is represented by a binary tree. A single input variable ($x$) and a split point on that numeric variable are interpreted by each node of the tree. The leaf nodes of the tree include an output variable ($y$), which is employed to make a straightforward prediction. The tree evaluates a new input started at the root node of the tree, whereas, a learned binary tree is regarded as a partitioning of the input space. The complexity of the decision tree algorithm is based on the number of splits in the tree. Pruning

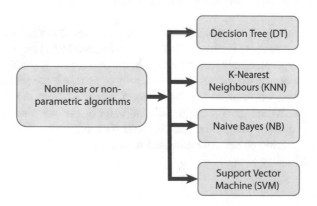

**Fig. 3.7** Nonlinear or non-parametric algorithms that can be used in sensor-based HAR

method is used to further lift performance. However, accuracy will suffer, if there is a non-linear relationship between predictors and outcome.

### 3.9.1.2  K-Nearest Neighbours (KNN)

KNN is one of the simplest algorithms as the entire training dataset is the model representation for KNN and predictions are made using the training dataset directly. No learning is required other than storing the entire dataset. Care must be taken about the consistency of training data and removal of erroneous and outlier data. When a new data point arrives, predictions are made by exploring through the full dataset for $K$ number of most similar instances (called as neighbors), and by compiling the output variable for those $K$ instances [69].

### 3.9.1.3  Naive Bayes (NB)

For the two-class (binary) and multiclass problem, Naive Bayes classification algorithm is used. In this methodology, the prediction probability is calculated by using three other statistical probability: likelihood probability, prior probability, and evidence probability. Class probability is also known as posterior probability. Here, likelihood probability denotes the probability of the instance such that it belongs to a specific class, whereas prior probability is the finding probability of that specific class among other classes and evidence probability is the finding probability of that specific instance among other instances.

$$p(y) = \frac{P(likelihood)P(prior)}{p(evidance)}$$

where, $y =$ predicted output.

### 3.9.1.4  Support Vector Machine (SVM)

Support Vector Machines can be explained by Maximal-Margin classifier [69]. In this case, an n-dimensional space is formed by the input variables of data. In SVM, a hyperplane or space is predict fot the best separation of the points in the input variable space by their class. The distance between the hyperplane and the nearest data points (support vectors) is called margin. Maximal-Margin hyperplane is the optimal line with the largest margin (perpendicular or vertical distance from the line to the most adjacent points) that can separate all the classes. The hyperplane is learned in the training phase. An optimization procedure is utilized to maximize the margin.

**Fig. 3.8** Parametric
algorithms that can be used
in sensor-based HAR

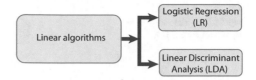

## 3.9.2 Linear Algorithms

Parametric algorithms interpret the function to a known form whereas, a parametric learning model compiles data with a set of parameters of fixed size (independent of the number of training examples). These models only care about the number of parameters needed no matter what amount of data is given as input. They are also called linear machine learning algorithms as the assumed functional form is a linear combination of the input variables. Parametric machine learning algorithms come with the benefits of easily interpreting results, very fast to learn from data, less training data requirement, and good working capability even if the fit to the data is not comprehensive. We have summarized the parametric algorithms that can be used for HAR in Fig. 3.8.

However, these algorithms have some limitations too, as these methods are constrained to the specified form, more suited to simpler problems, and unlikely to match the underlying mapping function in practice [69] (Fig. 3.8).

### 3.9.2.1 Logistic Regression (LR)

Logistic Regression is the go-to method for two-class values problems. The core of this method is based on logistic function or sigmoid function, which is an S-shaped curve that calculates the probability of the class by mapping the final output layer value from 0 to 1. Logistic regression is represented by the equation [69],

$$y = \frac{e^{B_0} + B_1 x}{1 + e^{B_0} + B_1 x}$$

where, $y$ = Predicted output, $B_0$ = Bias or intercept term, and $B_1$ = Coefficient for the single input value $(x)$.

A constant real-valued $B$ coefficient is associated with each column of input data that must be learned. Using maximum likelihood estimation, the coefficients are calculated from the training data, and the minimization algorithm is used to optimize the most appropriate values for the training data coefficients.

**Fig. 3.9** Ensemble
algorithms that can be used
in sensor-based HAR

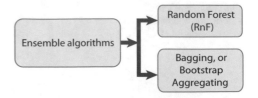

### 3.9.2.2    Linear Discriminant Analysis (LDA)

Linear Discriminant Analysis (LDA) works well when we have more than two classes
based classification problems. LDA models are represented by the mean and variance
of the variable for each class for a single input variable. In the case of multiple
variables, the same properties are calculated over the multivariate Gaussian namely
the means and covariance matrix. LDA model calculates the mean and variance for
each class simply assuming data to be Gaussian and the attributes to have the same
variance. Predictions are made by LDA by estimating the probability (using Bayes
theorem) that a new set of input belongs to each class. The class that gains the highest
probability is the output class and a prediction is made. The discriminant function is
calculated for each class and the class with the highest discriminant value makes the
output classification.

### 3.9.3    Ensemble Algorithms

Ensemble methods make use of multiple learning algorithms so that better predictive performance can be obtained than any of the constituent learning algorithm
alone [70].

A machine learning ensemble allows for a much more flexible formation to exist
among those alternatives, though it consists of only a solid finite set of alternative
models. Ensemble methods build a set of classifiers and then classify new data points
by exerting a (weighted) vote of their predictions. The original ensemble method
is Bayesian averaging, but more recent algorithms include error-correcting output
coding, Bagging, and boosting. We have summarized the ensemble algorithms that
can be used for HAR in Fig. 3.9.

### 3.9.4    Bagging or Bootstrap Aggregating

The algorithm is named as Bagging because it incorporates Bootstrapping and Aggre-
gation to create a single model ensemble. Multiple bootstrapped subsamples are
drawn based on a sample of data. On top every subsample, a Decision Tree is built.

After the creation of each subsample Decision Tree, an algorithm is used to aggregate over the Decision Trees to form the most effective predictor.

### 3.9.4.1  Random Forest (RnF)

Random Forest is the modern variation of classical Decision Tree algorithms. In the case of the Random Forest algorithm, the sub-trees are learned in such a way that the predictions from them have a less or weak correlation. The learning algorithm is restricted to a random sample of features of which to search. For classification problems, at each split point, $m$ number of features can be searched. A good default to find the value of $m$ is [69]

$$m = \sqrt{p}$$

where, $p =$ input variables number.

## 3.10  Overfitting and Underfitting Problem

If the estimator models the training data too well then it is known as overfitting. Overfitting happens when the model is aware of the information and noise in the training results. The effect of overfitting is that the model picks up and learns the noises of the training data as concepts. The problems are that a new data may not contain these noises. Therefore, these concepts may not be applied to any new data.

Underfitting means that the model is unable to fit the data well enough. An under-fitting machine learning model is not a proper model. Therefore, performance on training data is poor. We can say that an underfit model will not only perform poor for training data but also it will have poor generalization for other data.

## 3.11  Remedies of Overfitting and Underfitting Problem

Overfitting and underfitting both may lead to poor model results. In machine learning, overfitting is the most common problem. Evaluating the model and reporting results on the same dataset causes overfitting because the model will always make a more precise prediction of data that it has seen before. Underfit model is easy to detect and we need to try alternative machine learning models to prevent underfitting. To counter this overfitting, we have to test our model on unseen data. Therefore, to limit overfitting we can use the following techniques to evaluate our models as shown in Fig. 3.10.

**Fig. 3.10** Remedies of
overfitting problem

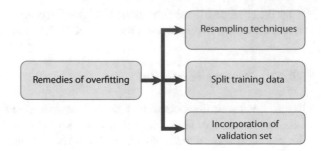

### 3.11.1　Resampling Techniques

K-fold cross-validation is the most famous and efficient resampling technique. In
this case, the training dataset is split into K subsets. Then the model is implemented
considering only set as the testing set and rests as the training set for K times. Finally,
we calculate the average accuracy of these K results.

### 3.11.2　Split Training Data

We can split some data from the training set or use completely different testing set to
predict the accuracy. In this case, we can estimate the learning models on the testing
dataset to get an opinion about the performance of the model on unseen data.

### 3.11.3　Incorporation of the Validation Set to Prevent
　　　　　Overfitting

In general, we need to split the dataset into two subsets. One of these subsets is used
for training the model, which is known as the train set, and the other is used for testing
the performance of the model, which is known is the test set. This is an important rule
of machine learning that the test data should be unseen by the model and after fitting
the model on the training we should evaluate the performance on the test data once.
If we tune the parameters of the model after evaluating the performance on test data
and evaluate the performance again, we will see that the performance is improving.
This is a case of overfitting. It implies that the test dataset is no longer invisible for
the model.

　　To solve this problem, the concept of validation dataset has come. The term
validation is related to validate something. It is defined as a subset of the full dataset,
which is isolated from training the model. It provides an estimate of the skill of the
model and allows to fine-tuning the various hyperparameters of the model. Every

type of experiment including model selection, neural network selection, number of layers and number of neuron selection, hyperparameter tuning, etc. are done based on the performance of the model on the validation set after fitting the model on the train set. Following adjustment of the parameters (i.e., tuning), the final appropriate model is used on the test data to assess the performance of the test dataset. This method solves the problem of overfitting on the test data and also provides scopes to tune the parameters of the model that eventually demonstrate superior performance. There are several ways to split the original dataset into the train, validation, and test set but as a thumb rule, it will be appropriate to split the main dataset into 70% as the training set, 10% as the validation set, and rest of the 20% data as the test set. So, we can summarize the train set, test set, and validation set by the following definitions:

- **Training set**: The training set is part of the data used to fit the model. In most of the cases, it covers all input and projected performance data. In the case of supervised machine learning, training data sets are labeled, whereas, an unlabeled training set is used for unsupervised machine learning techniques.
- **Validation set**: Validation set is defined as the sample of data utilized to give a balanced and fair analysis of a model to fit on the training dataset. The validation set tunes the hyperparameters of the model. Some authors put the results of validation set in their research papers though it may be unnecessary in most of the cases. As validation set is explored to tune and pick the well-suit hyperparameters, these results are less required to show. Mentioning the hyperparameters in the report is sufficient. The model does not learn from the validation set but the validation set can be effective for the model indirectly by providing an opportunity to improve the model without causing overfitting. If a dataset is tiny, then this subset and tuning can be avoided. In this scenario, there is a strong risk of overfitting where a model is learned and adjusted based on only the training sample. It will usually turn out that the model will not match the data for the real-world test as well as the data for the testing. Particularly when the size of the training data set is limited or when the amount of parameters in the model is high, the magnitude of this difference would likely be big. Cross-validation is a means of calculating the effect's scale. A distinction can be created between two forms of cross-validation: exhaustive and non-exhaustive cross-validation.

  - **Exhaustive cross-validation**: This is an approach of cross-validation that study and check all feasible means of splitting the initial sample into a training set and a validation. Leave-p-out and Leave-one-out cross-validation techniques fall under this category. Leave-p-out cross-validation (LpO CV) means utilizing p observations as the set for validation and the other observations as the set for training. This is replicated in every sense the original sample may be split on a validation set of $p$ findings and a training set [71]. If we consider a special case of Leave-p-out cross-validation with $p = 1$, then this is called Leave-one-out cross-validation [72].
  - **Non-exhaustive cross-validation**: Non-exhaustive cross-validation methods do not quantify all ways of initial sample splitting. Those methods are Leave-p-out approximations for cross-validation. k-fold cross-validation, Holdout method,

and Repeated random sub-sampling validation are examples from this category. In cross-validation k-fold, the initial sample is uniformly partitioned into sub-samples k of similar dimensions. Of the k subsamples, the validation data for evaluating the model is maintained as a single subsample, and the remaining k 1 subsample is used as training data. The cross-validation method is then replicated k-times, with each of the k subsamples being used as the validation data exactly once. Then, the values of k can be summed to provide one estimation [73]. We randomly allocate data points in the holdout process to two sets $D_0$ and $D_1$, which are typically named the training set and the test set, respectively. The scale of each of the sets is subjective because the evaluation package is usually smaller than the trial package. Dataset $D_0$ is used for training and evaluation of performance is done in $D_1$. Repeated random sub-sampling approach, also known as cross-validation of Monte Carlo, produces several random splits of the dataset into training and validation details [74, 75]. The model is fitted to the training data for each such split, and predictive performance is measured using validation data. Over the splits, the results are then averaged. The benefit of this approach (over k-fold cross-validation) is that the proportion of the split training/validation will not depend on the number of iterations.

Nested cross-validation is another variation of cross-validation that is needed when cross-validation is used simultaneously for the selection of the best collection of hyperparameters and error estimation (and generalization capability assessment). There are several variations of this. Each should discern at least two versions: k*l-fold cross-validation and k-fold cross-validation with validation and test set. k*l-fold cross-validation is a truly nested variant (used by [76] for example) which contains an outer loop of $k$ folds and an inner loop of $l$ folds. The entire collection of data is split into $k$-sets. k-fold cross-validation with validation and test set is a form of cross-validation k*l-fold while running $l = k - 1$.

- **Test set**: Test set is defined as the subset of data that is explored to implement in the final model—to assess the performance of the model. These are mentioned as the final results of a model under a dataset. This is considered as a unbiased assessment of the model. Test set is used only when the model has been completely trained. We also need to be careful so that the test set contains sampled data from different classes and we need to make sure there is no class imbalance.

### 3.11.4  Overfitting Related to Overlapping Sliding Window and Cross-Validation with Random Splitting

If we use the sliding window with some percentage of overlap and K-fold cross-validation with the random splitting of data, it can cause overfitting. While splitting randomly, data from one window may be put on the train set, whereas data from the most adjacent window may be put into the test set. As we are using a sliding window

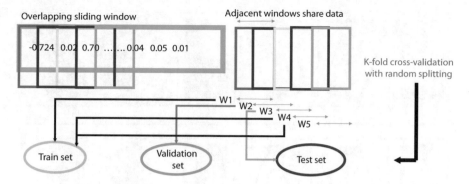

**Fig. 3.11** Overfitting due to the simultaneous use of overlapping sliding window and cross-validation with random splitting

with overlapping, it means that adjacent windows share some percentages of data. It will cause similar data to appear on both test and train sets, which will expose some of the data from the test set while fitting the model as they also appear in the train set. This will cause overfitting. This situation has been shown in Fig. 3.11.

We can solve this problem by using a non-overlapping sliding window or by using K-fold cross-validation with the temporal splitting of data while using an overlapping sliding window. In this case, we need to put the data into train, validation, and test set in a temporal manner. For example, we can put the first 60% window data in the training set, the next 20% data in the validation set, and finally the rest of the 20% window data in the test set. We have called this splitting as temporal splitting as the splitting of data maintains temporal order. Thus, the splitting of data in three sets from the overlapping windows should not be random. The temporal splitting of data will have a temporal order of windows, which will prevent the data from the adjacent windows to appear in train and test sets.

### 3.11.5  Overfitting Because of Orientation-Dependent Model

In general, most of the datasets are created using a fixed oriented smartphone or wearable device. A model which is trained on these sensor data performs well on the test data also, as the user carry a smartphone or wearable device in the same orientation trained by the model. These types of models are called orientation-dependent model, which is overfitting because of a specific orientation. These models will show poor performance in real-time where users may carry the smartphone or wearable device in different orientations as shown in Fig. 3.12. To prevent this overfitting the model should be trained such a way so that it can offer the same performance even if the orientation changes in real-time.

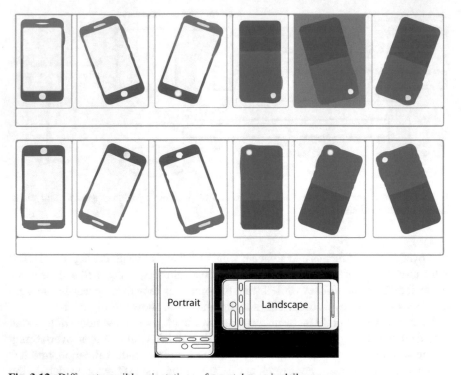

**Fig. 3.12** Different possible orientations of smartphones in daily usage

### 3.11.6  Overfitting Because of Position-Dependent Model

In previous research works, it was a common practice to make the datasets by keeping the smartphone or wearable sensors in fixed body positions like in the shirt pocket, or pant pocket, or hand, and so on. In previous research works, we have found many considerable body positions to place the wearable sensor device, e.g.,

- Waist [24, 77–85],
- Chest [67, 83, 85–90],
- Wrist [24, 81, 82, 85, 88, 90, 91],
- Upper arm [24, 92],
- Head [92, 93],
- Thigh [24, 77, 86, 88, 91],
- Leg [91],
- Ear [12],
- Ankle [82, 83, 85], etc.

In Fig. 3.13, we have summarized the possible various positions of wearable sensor devices and smartphone.

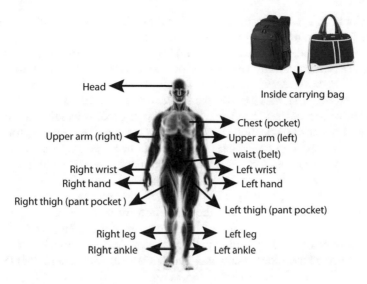

**Fig. 3.13** Different possible positions of smartphones and wearable devices

This practice puts a constraint on the user behavior, and therefore, a model can be position-dependent. And it will cause overfitting as the model will only show good performance for a specific position. In real life, most of the users do not like to keep the smartphone or wearable devices in any specific position. If this happens, the position-dependent model will show poor performance. This is why it is necessary to create a dataset by keeping the smartphone in different positions, and the researchers should focus to build a position-independent model. The performance of that model can be examined by training the model using specific body sensor data (e.g., shirt pocket and pant pocket), and testing the model using other body sensor data (like the hand). In this way, we can build a position-independent model for real-life cases to prevent overfitting.

## 3.12  Some Analysis on Behavior Identification Using Traditional Approaches For Pattern Recognition

In this section, we have provided a comparative study of some of the standard databases of the classification methods in previous studies. We have also compared numerous feature selection processes and classification accuracies. In [94], they built a method by using a waist-mounted accelerometer to collect data from 6 subjects which included 12 daily activities. This work has proposed to deliver the majority of signal processing onboard the wearable unit employing embedded intelligence and implementation of a real-time classification system. They reached an average accuracy of 90.8%, while postural orientation identification was 94.1% accurate and probable falls were observed with 95.6% accuracy.

On the other hand, one-dimensional (1D) Haar-like filtering techniques have been introduced as a unique feature extraction method suitable for 3D with lower computational cost in the paper [95]. Their new approach reached 93.91% precision in case of activity recognition, thus growing the expense of computation to 21.22% relative to a few prior approaches. The work [96] has attempted to identify the important sensors and better classification process by using supervised learning for activity recognition. A two-step prediction technique has been used by them. In the first step, they have used binary classifiers to distinguish between the null class and the rest of the classes. In the second step, a multi-class classifier has been exploited to predict the exact class. In [97], a Kernel Principal Component Analysis (KPCA) has further processed features after the extraction for better accuracy. Linear Discriminant Analysis (LDA) was used to render them sturdier. They ended up using a Deep Belief Network (DBN) to train the features. They noticed 89.61% accuracy that outperformed traditional multiclass Vector Support Machine (SVM) (82.02%) and Artificial Neural Network (ANN) (65.31%).

Summary of features, classification methods, proposed solutions and accuracies of previous works have been discussed in Table 3.2 for HASC2010 Corpus, UCI HAR, UCI HAPT, UCI batteryless wearable sensor, UniMiB SHAR, and WARD dataset.

## 3.13  Conclusion

This chapter presents a basic outline of the methodology of sensor-based HAR including feature extraction, feature selection, feature normalization, conventional machine learning techniques, and problems of overfitting. Hand-crafted features have been discussed in detail including time domain, frequency domain, wavelet, and heuristic features. To balance the trade-off between performance and computational cost, we have discussed some prominent feature selection and feature normalization techniques. Linear and non-linear algorithms of machine learning have been considered in this chapter to classify activity data. Moreover, the problems and causes of overfitting and underfitting problems have been addressed with remedies. Finally, an analysis has been provided comparing the methods of previous research works on some benchmark datasets using these methods.

## 3.14  Think Further

1. What are the importance of extracting features from sensor data?
2. What are the differences between time domain and frequency domain features?
3. Provide some examples of time domain features.

**Table 3.2** Comparative analysis of existing works on some benchmark datasets

| Dataset | Paper | Features | Method | Comment |
|---|---|---|---|---|
| UCI HAR [99] | [98] | 17 | Multiclass SVM. Accuracy: 89.3% | They used fixed point arithmetic, which is energy-efficient. |
| | [100] | 561 | One-Vs-One SVM and KNN. 96.4% accuracy. | They proposed a majority voting system. |
| | [101] | 561 | Sparse kernelized Learning Vector Quantization (LVQ) model. 96.23% accuracy. | They proposed metric adaptation with 1 prototype vector for LVQ. |
| | [102] | 561 | Novel confidence based boosting algorithm. 94.33% accuracy. | They exploited confidence information from weak learner. |
| UCI HAPT [104] | [103] | 561 | SVM with probability and discrete filtering. 95.24% and 96.66% accuracy. | They estimated the probability of each activity to be on each class. |
| HASC Corpus [106] | [105] | 12 | Random Forest (accuracy: 75%) with EM+DENSE and EM+SPARSE method. | They improved activity recognition with incorrect segments. |
| UCI BWS [108] | [107] | 46 | CRF classifier. Precision: 85.1% for room-1 and 84.9% for room-2. | They proposed a bed-egress movement detection framework. |
| UniMiB SHAR [109] | [109] | 2 | 4 classifiers KNN- 82.86%, SVM- 98.71%, ANN- 72.13% and RnF- 88.41%. | They monitored different falls and daily activities. |
| WARD SHAR [110] | [110] | 40 | KNN- 90.5% with 1–5 sensors. | They used a framework with majority voting and Distributed Sparsity Classifier (DSC). |

4. What are the purposes of using mean, standard deviation, median absolute deviation, etc. features for human activity recognition?
5. How can we extract frequency domain feature?
6. What are the impacts of frequency domain feature to distinguish human activities?
7. Provide some examples of frequency domain features.
8. What are the importance of using wavelet analysis?
9. Provide some examples of wavelet analysis-based features.
10. What are the importance of heuristic features?
11. Provide some examples of heuristic features.
12. State some importance of feature selection and dimensionality reduction techniques.
13. Mention some process of feature selection and dimensionality reduction.
14. How to select appropriate feature selection and dimensionality reduction techniques in different cases?
15. What are the importance of feature normalization?
16. What are the basic feature normalization processes?
17. Why most of the human activity classifications can be done using supervised machine learning techniques?
18. Mention some nonlinear machine learning techniques.
19. Mention some linear machine learning algorithms.
20. What are overfitting and underfitting problem?
21. State some remedies of overfitting and underfitting problem.
22. How can overfitting occur in the case of overlapping sliding window and cross-validation with random splitting?
23. How can a orientation-dependent model cause overfitting in real-time cases?
24. Which type of conventional pattern recognition methods have been used in earlier research works?
25. Mention some benefits of conventional pattern recognition approaches.
26. Mention some drawbacks and limitations of earlier research works.

# References

1. Antar, A.D., Ahad, M.A.R., Shahid, O.: Vision-based action understanding for assistive healthcare: a short review. IEEE CVPR Workshop (2019)
2. Ahad, M.A.R.: Vision and sensor based human activity recognition: challenges ahead (2020)
3. Antar, A.D., Ahmed, M., Ahad, M.A.R.: Challenges in sensor-based human activity recognition and a comparative analysis of benchmark datasets: a review. In: 2019 Joint 8th International Conference on Informatics, Electronics & Vision (ICIEV) and 2019 3rd International Conference on Imaging, Vision & Pattern Recognition (icIVPR), pp. 134–139. IEEE, Washington, D.C. (2019)
4. Ahad, M.A.R.: Motion History Images for Action Recognition and Understanding. Springer Science & Business Media, Berlin (2012)

5. Ahad, M.A.R.: Computer Vision and Action Recognition: A Guide for Image Processing and Computer Vision Community for Action Understanding, vol. 5. Springer Science & Business Media, Berlin (2011)
6. Rafiuddin, N., Khan, Y.U., Farooq, O.: Feature extraction and classification of eeg for automatic seizure detection. In: Multimedia, Signal Processing and Communication Technologies (IMPACT), 2011 International Conference on, pp. 184–187. IEEE, Aligarh (2011)
7. Veltink, P.H., Bussmann, H.B.J., de Vries, W., Martens, W.L.J., van Lummel, R.C.: Detection of static and dynamic activities using uniaxial accelerometers. IEEE Trans. Rehabil. Eng. **4**, 375–385 (1996)
8. Hossain, T., Goto, H., Ahad, M.A.R., Inoue, S.: A study on sensor-based activity recognition having missing data. In: 2018 Joint 7th International Conference on Informatics, Electronics & Vision (ICIEV) and 2018 2nd International Conference on Imaging, Vision & Pattern Recognition (icIVPR), pp. 556–561. IEEE, Kitakyushu (2018)
9. Comparative study on classifying human activities with miniature inertial and magnetic sensors. Patt. Recogn. **43**(10), 3605–3620 (2010)
10. Saha, S.S., Rahman, S., Rasna, M.J., Zahid, T.B., Mahfuzul Islam, A.K.M., Ahad, M.A.R.: Feature extraction, performance analysis and system design using the du mobility dataset. IEEE Access **6**, 44776–44786 (2018)
11. Rasna, M.J., Hossain, T., Inoue, S., Sha, S.S., Rahman, S., Ahad, M.A.R.: Supervised and neural classifiers for locomotion analysis. In: 2018 ACM International Joint Conference on Pervasive and Ubiquitous Computing and the 2018 International Symposium on Wearable Computers (UbiComp/ISWC) (2018)
12. Hossain, T., Islam, M.S., Ahad, M.A.R., Inoue, S.: Human activity recognition using earable device. In: Proceedings of the 2019 ACM International Joint Conference on Pervasive and Ubiquitous Computing and Proceedings of the 2019 ACM International Symposium on Wearable Computers, pp. 81–84. ACM, New York (2019)
13. Bouten, C.V.C., Koekkoek, K.T.M., Verduin, M., Kodde, R., Janssen, J.D.: A triaxial accelerometer and portable data processing unit for the assessment of daily physical activity. IEEE Trans. Biomed. Eng. **44**(3), 136–147 (1997). March
14. Antar, A.D., Ahmed, M., Hossain, T., Muramatsu, D., Makihara, Y., Inoue, S., Yagi, Y., Ahad, M.A.R., Ngo, T.T.: Wearable sensor-based gait analysis for age and gender estimation (2020)
15. Tazin, T., Hossain, T., Ahad, M.A.R., Inoue, S.: Activity recognition by using lorawan sensor. In: 2018 ACM International Joint Conference on Pervasive and Ubiquitous Computing and the 2018 International Symposium on Wearable Computers (UbiComp/ISWC) (2018)
16. Pirttikangas, S., Fujinami, K., Nakajima, T.: Feature selection and activity recognition from wearable sensors. In: Youn, H.Y., Kim, M., Morikawa, H., (eds.) Ubiquitous Computing Systems, pp. 516–527. Springer, Berlin, Heidelberg (2006)
17. Attal, F., Mohammed, S., Dedabrishvili, M., Chamroukhi, F., Oukhellou, L., Amirat, Y.: Physical human activity recognition using wearable sensors. Sensors **15**(12), 31314–31338 (2015)
18. Farringdon, J., Moore, A.J., Tilbury, N., Church, J., Biemond, P.D.: Wearable sensor badge and sensor jacket for context awareness. In: Digest of Papers. Third International Symposium on Wearable Computers, pp. 107–113 (1999)
19. Oliphant, T.: Guide to Numpy. Tregol Publishing (2006)
20. Sekine, M., Tamura, T., Fujimoto, T., Fukui, Y.: Classification of walking pattern using acceleration waveform in elderly people. In: Engineering in Medicine and Biology Society, 2000. Proceedings of the 22nd Annual International Conference of the IEEE, vol. 2, pp. 1356–1359. IEEE (2000)
21. Randell, C., Muller, H.: Context awareness by analysing accelerometer data. In: Wearable Computers, The Fourth International Symposium on, pp. 175–176. IEEE (2000)
22. Ravi, N., Dandekar, N., Mysore, P., Littman, M.L.: Activity recognition from accelerometer data. In: Proceedings of the Seventeenth Conference on Innovative Applications of Artificial Intelligence (IAAI) (2005)
23. Machado, I.P.: Human activity data discovery based on accelerometry (2013)

24. Bao, L., Intille, S.S.: Activity recognition from user-annotated acceleration data. In: International Conference on Pervasive Computing, pp. 1–17. Springer, Berlin (2004)
25. Box, G.E.P., Jenkins, G.M.: Time Series Analysis: Forecasting and Control (1976)
26. Schmidt, A., Van Laerhoven, K.: How to build smart appliances? IEEE Pers. Commun. **8**(4), 66–71 (2001)
27. Huile, X., Liu, J., Haibo, H., Zhang, Y.: Wearable sensor-based human activity recognition method with multi-features extracted from hilbert-huang transform. Sensors **16**(12), 2048 (2016)
28. Hamäläinen, W., Järvinen, M., Martiskainen, P., Mononen, J.: Jerk-based feature extraction for robust activity recognition from acceleration data. In: 2011 11th International Conference on Intelligent Systems Design and Applications, pp. 831–836 (2011)
29. Xueshan, Y., Xiaozhai, Q., Lee, G.C., Tong, M., Jinming,C.: Jerk and jerk sensor 06 (2018)
30. Hamäläinen, W., Järvinen, M., Martiskainen, P., Mononen, J.: Jerk-based feature extraction for robust activity recognition from acceleration data. In: Intelligent Systems Design and Applications (ISDA), 2011 11th International Conference on, pp. 831–836. IEEE, Piscataway, NJ (2011)
31. Foerster, F., Smeja, M., Fahrenberg, J.: Detection of posture and motion by accelerometry: a validation study in ambulatory monitoring. Comput. Hum. Behav. **15**, 571–583 (1999)
32. Preece, S.J., Goulermas, J.Y., Kenney, L.P.J., Howard, D., Meijer, K., Crompton, R.: Physiological Measurement
33. Englehart, K., Hudgins, B., Parker, P., Stevenson, M.: Time-frequency representation for classification of the transient myoelectric signal. In: Proceedings of the 20th Annual International Conference of the IEEE Engineering in Medicine and Biology Society. vol. 20 Biomedical Engineering Towards the Year 2000 and Beyond (Cat. No.98CH36286), vol. 5, pp. 2627–2630 (1998)
34. Bao, L., Intille, S.S.: Activity recognition from user-annotated acceleration data. Pervasive Computing, pp. 158–175. Springer, Berlin/Heidelberg (2004)
35. Nyan, M.N., Tay, F.E.H., Seah, K.H.W., Sitoh, Y.Y.: Classification of gait patterns in the time-frequency domain. J. Biomech. **39**(14), 2647–2656 (2006)
36. Najafi, B., Aminian, K., Paraschiv-Ionescu, A., Loew, F., Bula, C.J., Robert, P.: Ambulatory system for human motion analysis using a kinematic sensor: monitoring of daily physical activity in the elderly. IEEE Trans. Biomed. Eng. **50**(6), 711–723 (2003)
37. Mathie, M.J., Coster, A.C.F., Lovell, N.H., Celler, B.G.: Accelerometry: providing an integrated, practical method for long-term, ambulatory monitoring of human movement. Physiological measurement **25**(2), R1 (2004)
38. Makikawa, M., Iizumi, H.: Development of an ambulatory physical activity memory device and its application for the categorization of actions in daily life. Medinfo **8**, 747–750 (1995)
39. Bussmann, J.B.J., Martens, W.L.J., Tulen, J.H.M., Schasfoort, F.C., Van Den Berg-Emons, H.J.G., Stam, H.J.: Measuring daily behavior using ambulatory accelerometry: the activity monitor. Behav. Res. Methods Instrum. Comput. **33**(3), 349–356 (2001)
40. Veltink, P.H., Bussmann, H.J., Vries, W.D., Martens, W.J., Van Lummel, R.C.: Detection of static and dynamic activities using uniaxial accelerometers. IEEE Trans. Rehabil. Eng. **4**(4), 375–385 (1996)
41. Aminian, K., Ph Robert, E.E., Buchser, B.R., Hayoz, D., Depairon, M.: Physical activity monitoring based on accelerometry: validation and comparison with video observation. Med. Biol. Eng. Comput. **37**(3), 304–308 (1999)
42. Mladenić, D.: Feature selection for dimensionality reduction. In: International Statistical and Optimization Perspectives Workshop Subspace, Latent Structure and Feature Selection, pp. 84–102. Springer, Berlin (2005)
43. Peng, H., Long, F., Ding, C.: Feature selection based on mutual information criteria of max-dependency, max-relevance, and min-redundancy. IEEE Trans. Pattern Anal. Mach. Intell. **27**(8), 1226–1238 (2005)
44. Jatoba, L.C., Grossmann, U., Kunze, C., Ottenbacher, J., Stork, W.: Context-aware mobile health monitoring: evaluation of different pattern recognition methods for classification of

physical activity. In: Engineering in Medicine and Biology Society, 2008. EMBS 2008. 30th Annual International Conference of the IEEE, pp. 5250–5253. IEEE, Piscataway, NJ (2008)

45. Maurer, U., Smailagic, A., Siewiorek, D.P., Deisher, M.: Activity recognition and monitoring using multiple sensors on different body positions. In: Wearable and Implantable Body Sensor Networks, 2006. BSN 2006. International Workshop on, pp. 4. IEEE, Cambridge, MA (2006)

46. Maurer, U., Smailagic, A., Siewiorek, D.P., Deisher, M.: Activity recognition and monitoring using multiple sensors on different body positions. In: Wearable and Implantable Body Sensor Networks, 2006. BSN 2006. International Workshop on, pp. 4–pp. IEEE, Cambridge, MA (2006)

47. Hira, Z.M., Gillies, D.F.: A review of feature selection and feature extraction methods applied on microarray data. Adv. Bioinform. **2015**, 13 (2015)

48. Guyon, I., Elisseeff, A.: An introduction to variable and feature selection. J. Mach. Learn. Res. **3**, 1157–1182 (2003)

49. Miao, J., Niu, L.: A survey on feature selection. Proc. Comput. Sci. **91**, 919–926 (2016)

50. Forman, G.: An extensive empirical study of feature selection metrics for text classification. J. Mach. Learn. Res. **3**, 1289–1305 (2003)

51. Chen, X.-W., Jeong, J.C.: Enhanced recursive feature elimination. In: Sixth International Conference on Machine Learning and Applications (ICMLA 2007), pp. 429–435. IEEE, Cincinnati, OH (2007)

52. Talenti, L., Luck, M., Yartseva, A., Argy, N., Houzé, S., Damon, C.: L1 logistic regression as a feature selection step for training stable classification trees for the prediction of severity criteria in imported malaria. arXiv:1511.06663 (2015)

53. van der Maaten, L., Hinton, G.: Visualizing data using t-sne. J. Mach. Learn. Res. **9**, 2579–2605 (2000)

54. McInnes, L., Healy, J., Melville, J.: Umap: Uniform manifold approximation and projection for dimension reduction. arXiv:1802.03426 (2018)

55. Ali, M.U., Ahmed, S., Ferzund, J., Mehmood, A., Rehman, A.: Using pca and factor analysis for dimensionality reduction of bio-informatics data. arXiv:1707.07189 (2017)

56. Shi, H., Yin, B., Zhang, X., Kang, Y., Lei, Y.: A landmark selection method for l-isomap based on greedy algorithm and its application. In: 2015 54th IEEE Conference on Decision and Control (CDC), pp. 7371–7376. IEEE, Oskara (2015)

57. Hall, M.A.: Correlation-based feature selection for machine learning (1999)

58. Baranauskas, J.A., Netto, O.P., Nozawa, S.R., Macedo, A.A.: A tree-based algorithm for attribute selection. Appl. Intell. **48**(4), 821–833 (2018)

59. Mantyjarvi, J., Himberg, J., Seppanen, T.: Recognizing human motion with multiple acceleration sensors. In: Systems, Man, and Cybernetics, 2001 IEEE International Conference on, vol. 2, pp. 747–752. IEEE, Tucson, AZ (2001)

60. He, Z., Jin, L.: Activity recognition from acceleration data based on discrete consine transform and svm. In: Systems, Man and Cybernetics, 2009. SMC 2009. IEEE International Conference on, pp. 5041–5044. IEEE, San Antonio (2009)

61. He, Z.-Y., Jin, L.-W.: Activity recognition from acceleration data using ar model representation and svm. In: Machine Learning and Cybernetics, 2008 International Conference on, vol. 4, pp. 2245–2250. IEEE, Kunming (2008)

62. Khalid, S., Khalil, T., Nasreen, S.: A survey of feature selection and feature extraction techniques in machine learning. In: 2014 Science and Information Conference, pp. 372–378. IEEE, London (2014)

63. Janecek, A., Gansterer, W., Demel, M., Ecker, G.: On the relationship between feature selection and classification accuracy. In: New challenges for feature selection in data mining and knowledge discovery, pp. 90–105 (2008)

64. Joel, G.: Data science from scratch: first principles with python (2015)

65. Stikic, M., Van Laerhoven, K., Schiele, B.: Exploring semi-supervised and active learning for activity recognition. In: Wearable computers, 2008. ISWC 2008. 12th IEEE international symposium on, pp. 81–88. IEEE (2008)

66. Machado, I.P.: Human activity data discovery based on accelerometry. PhD thesis, Faculdade de Ciências e Tecnologia (2013)
67. Khan, A.M., Lee, Y.-K., Lee, S.Y., Kim, T.-S.: A triaxial accelerometer-based physical-activity recognition via augmented-signal features and a hierarchical recognizer. IEEE Trans. Inf. Technol. Biomed. **14**(5), 1166–1172 (2010)
68. The aware home.: http://awarehome.imtc.gatech.edu Accessed 02 March 2019
69. The Domus Laboratory.: http://domus.usherbrooke.ca Accessed 02 March 2019
70. The iDorm Project.: http://cswww.essex.ac.uk/iieg/idorm.htm. Accessed 02 March 2019
71. Celisse, A., et al.: Optimal cross-validation in density estimation with the $l\{2\}$ -loss. Ann. Stat. **42**(5), 1879–1910 (2014)
72. Molinaro, A.M., Simon, R., Pfeiffer, R.M.: Prediction error estimation: a comparison of resampling methods. Bioinformatics **21**(15), 3301–3307 (2005)
73. McLachlan, G.J., Do, K.-A., Ambroise, C.: Analyzing Microarray Gene Expression Data, vol. 422. Wiley, Hoboken (2005)
74. Dubitzky, W., Granzow, M., Berrar, D.P.: Fundamentals of Data Mining in Genomics and Proteomics. Springer Science & Business Media, Berlin (2007)
75. Landry, C.: Market transfers of water for environmental protection in the western united states. Water Policy **1**(5), 457–469 (1998)
76. Cawley, G.C., Talbot, N.L.C.: Cawley2010over, on over-fitting in model selection and subsequent selection bias in performance evaluation. J. Mach. Learn. Res. **11**, 2079–2107 (2010). JMLR. Org
77. Yeoh, W.-S., Pek, I., Yong, Y.-H., Chen, X., Waluyo, A.B.: Ambulatory monitoring of human posture and walking speed using wearable accelerometer sensors. In: 2008 30th Annual International Conference of the IEEE Engineering in Medicine and Biology Society, pp. 5184–5187. IEEE (2008)
78. Mathie, M.J., Coster, A.C.F., Lovell, N.H., Celler, B.G.: Detection of daily physical activities using a triaxial accelerometer. Med. Biol. Eng. Comput. **41**(3), 296–301 (2003)
79. Baek, J., Lee, G., Park, W., Yun, B.-J.: Accelerometer signal processing for user activity detection. In: International Conference on Knowledge-Based and Intelligent Information and Engineering Systems, pp. 610–617. Springer, Berlin (2004)
80. Meijer, G.A.L., Westerterp, K.R., Verhoeven, F.M.H., Koper, H.B.M., ten Hoor, F.: Methods to assess physical activity with special reference to motion sensors and accelerometers. IEEE Trans. Biomed. Eng. **38**(3), 221–229 (1991)
81. Maurer, U., Smailagic, A., Siewiorek, D.P., Deisher, M.: Activity recognition and monitoring using multiple sensors on different body positions. Technical Report, Carnegie-Mellon Univ Pittsburgh PA school of Computer Science (2006)
82. Tamura, T., Fujimoto, T., Muramoto, H., Huang, J., Sakaki, H., Togawa, T.: The design of an ambulatory physical activity monitor and it application to the daily activity of the elderly. In: Proceedings of 17th International Conference of the Engineering in Medicine and Biology Society, vol. 2, pp. 1591–1592. IEEE (1995)
83. Mehmood, A., Raza, A., Nadeem, A., Saeed, U.: Study of multi-classification of advanced daily life activities on shimmer sensor dataset. Int. J. Commun. Netw. Inf. Secur. **8**(2), 86 (2016)
84. Gupta, P., Dallas, T.: Feature selection and activity recognition system using a single triaxial accelerometer. IEEE Trans. Biomed. Eng. **61**(6), 1780–1786 (2014)
85. Leutheuser, H., Schuldhaus, D., Eskofier, B.M.: Hierarchical, multi-sensor based classification of daily life activities: comparison with state-of-the-art algorithms using a benchmark dataset. PLoS One **8**(10), e75196 (2013)
86. Lyons, G.M., Culhane, K.M., Hilton, D., Grace, P.A., Lyons, D.: A description of an accelerometer-based mobility monitoring technique. Med. Eng. Phys. **27**(6), 497–504 (2005)
87. Lara, O.D., Pérez, A.J., Labrador, M.A., Posada, J.D.: Centinela: a human activity recognition system based on acceleration and vital sign data. Pervas. Mob. Comput. **8**(5), 717–729 (2012)
88. McGlynn, D., Madden, M.G.: An ensemble dynamic time warping classifier with application to activity recognition. In: International Conference on Innovative Techniques and Applications of Artificial Intelligence, pp. 339–352. Springer, Berlin (2010)

89. Ermes, M., Parkka, J., Cluitmans, L.: Advancing from offline to online activity recognition with wearable sensors. In: 2008 30th Annual International Conference of the IEEE Engineering in Medicine and Biology Society, pp. 4451–4454. IEEE (2008)

90. Parkka, J., Ermes, M., Korpipaa, P., Mantyjarvi, J., Peltola, J., Korhonen, I.: Activity classification using realistic data from wearable sensors. IEEE Trans. Inf. Technol. Biomed. **10**(1), 119–128 (2006)

91. Ahmadi, A., Mitchell, E., Richter, C., Destelle, F., Gowing, M., O'Connor, N.E., Moran, K.: Toward automatic activity classification and movement assessment during a sports training session. IEEE Internet Things J. **2**(1), 23–32 (2014)

92. Min, J.-K., Cho, S.-B.: Activity recognition based on wearable sensors using selection/fusion hybrid ensemble. In: 2011 IEEE International Conference on Systems, Man, and Cybernetics, pp. 1319–1324. IEEE, Anchorage (2011)

93. Menz, H.B., Lord, S.R., Fitzpatrick, R.C.: Age-related differences in walking stability. Age Age. **32**(2), 137–142 (2003)

94. Karantonis, D.M., Narayanan, M.R., Mathie, M., Lovell, N.H., Celler, B.G.: Implementation of a real-time human movement classifier using a triaxial accelerometer for ambulatory monitoring. IEEE Trans. Inf. Technol. Biomed. **10**(1), 156–167 (2006)

95. Hanai, Y., Nishimura, J., Kuroda, T.: Haar-like filtering for human activity recognition using 3d accelerometer. In: Digital Signal Processing Workshop and 5th IEEE Signal Processing Education Workshop, 2009. DSP/SPE 2009. IEEE 13th, pp. 675–678. IEEE, Marco Island, FL (2009)

96. Botros, M., Heskes, T., de Vries, I.A.P.: Supervised learning in human activity recognition based on multimodal body sensing (2017)

97. Hassan, M.M., Uddin, M.Z., Mohamed, A., Almogren, A.: A robust human activity recognition system using smartphone sensors and deep learning. Future Gener. Comput. Syste. **81**, 307–313 (2018)

98. Anguita, D., Ghio, A., Oneto, L., Parra, F.X.L., Ortiz, J.L.R.: Energy efficient smartphone-based activity recognition using fixed-point arithmetic. J. Univ. Comput. Sci. **19**(9), 1295–1314 (2013)

99. Anguita, D., Ghio, A., Oneto, L., Parra, X., Reyes-Ortiz, J.L.: A public domain dataset for human activity recognition using smartphones. In: ESANN (2013)

100. Romera-Paredes, B., Aung, M.S.H., Bianchi-Berthouze, N.: A one-vs-one classifier ensemble with majority voting for activity recognition. In: ESANN 2013 proceedings, 21st European Symposium on Artificial Neural Networks, Computational Intelligence and Machine Learning, pp. 443–448 (2013)

101. Kästner, M., Strickert, M., Villmann, T., Mittweida, S.-G.: A sparse kernelized matrix learning vector quantization model for human activity recognition. In: ESANN (2013)

102. Reiss, A., Hendeby, G., Stricker, D.: A competitive approach for human activity recognition on smartphones. In: European Symposium on Artificial Neural Networks, Computational Intelligence and Machine Learning (ESANN 2013), 24-26 April, Bruges, Belgium, pp. 455–460 (2013)

103. Reyes-Ortiz, J.-L., Oneto, L., Ghio, A., Samá, A., Anguita, D., Parra, X.: Human activity recognition on smartphones with awareness of basic activities and postural transitions. In: International Conference on Artificial Neural Networks, pp. 177–184. Springer, Berlin (2014)

104. Reyes-Ortiz, J.-L., Oneto, L., Samà, A., Parra, X., Anguita, D.: Transition-aware human activity recognition using smartphones. Neurocomputing **171**, 754–767 (2016)

105. Toda, T., Inoue, S., Ueda, N.: Mobile activity recognition through training labels with inaccurate activity segments. In: Proceedings of the 13th International Conference on Mobile and Ubiquitous Systems: Computing, Networking and Services, pp. 57–64. ACM, New York (2016)

106. Hasc2010 corpus.: http://hasc.jp. Accessed 27 March 2019

107. Wickramasinghe, A., Ranasinghe, D.C., Fumeaux, C., Hill, K.D., Visvanathan, R.: Sequence learning with passive rfid sensors for real-time bed-egress recognition in older people. IEEE J. Biomed. Health Inform. **21**(4), 917–929 (2017)

108. Torres, R.L.S., Ranasinghe, D.C., Shi, Q., Sample, A.P.: Sensor enabled wearable rfid technology for mitigating the risk of falls near beds. In: RFID (RFID), 2013 IEEE International Conference on, pp. 191–198. IEEE, Penang (2013)
109. Micucci, D., Mobilio, M., Napoletano, P.: Unimib shar: a dataset for human activity recognition using acceleration data from smartphones. Appl. Sci. **7**(10), 1101 (2017)
110. Yang, A.Y., Jafari, R., Sastry, S.S., Bajcsy, R.: Distributed recognition of human actions using wearable motion sensor networks. J. Ambient Intell. Smart Environ. **1**(2), 103–115 (2009)

# Chapter 4
# Human Activity Recognition: Data Collection and Design Issues

**Abstract** Sensor-based Human Activity Recognition (HAR) has been explored by many research communities and industries for various applications. In the earlier chapters, we have presented methodologies to accomplish human activity recognition, pre-processing steps of raw data from sensors, segmentation of these data using various windowing approaches, feature extraction approaches, feature normalization schemes, sub-space methods for dimensional reduction and related issues. In this chapter, we present important challenges in activity recognition, data collection protocols, and design issues. This chapter also represents the basic requirement of training data, environmental set up for data collection, sensor requirement, sensor position, and energy consumption issues.

## 4.1 Human Activity Recognition: Data Collection

Sensor-based Human Activity Recognition (HAR) has been explored by many research communities and industries for various applications—along with various challenges ahead to deal with [1–8]. Human Activity Recognition (HAR) algorithms can be assessed on the footing of the complexity of the activities they acknowledge in the field of sensor-based activity recognition. The primary complications depend on various factors including the number of activities, types of activities, choice of sensors, energy expenditure, obtrusiveness, and data collection protocols.

### 4.1.1 Complications of the Activities

Recognition algorithms can be assessed based on the complexity of the activities they recognize in the field of sensor-based recognition of daily human activities. The complexity of the activities and actions can change and depends on various determinants including the number of activities, the types of activities, and the complexity of the training data obtained for those activities.

© Springer Nature Switzerland AG 2021
M. A. R. Ahad et al., *IoT Sensor-Based Activity Recognition*, Intelligent Systems Reference Library 173, https://doi.org/10.1007/978-3-030-51379-5_4

There are various benchmark datasets where the number of the activities vary from 6 to 10 classes usually. Some of the common activity classes are walking, running, sitting, stair up, stair down, etc. The variabilities of classes make it difficult to form a single model to work on many classes. Moreover, how these classes are extracted and what kind of sensors are explored, are crucial issues.

#### 4.1.1.1   Number of Activities

The complications of how precisely a human activity recognition system can identify a specific activity increase with a distinct and extensive set of activities people perform. However, the recognition of a large set of activities is difficult than the recognition of a smaller set of activities. The logic for this can be connected to the fact that as activities rise in number, general classifiers need to distinguish these classes among a larger set of activities. This is why the recognition rate with optimum performance is difficult to achieve.

#### 4.1.1.2   Patterns of Activities

There are three basic patterns of (ambulation) activities in total [3] (as shown in Fig. 4.1):

- Static activities,
- Dynamic activities, and
- Activities with postural transitions.

Static activities such as lying, sitting, and standing, are easier to identify than the periodic activities, such as running, walking, jogging, etc. based on any complex movement and actions. However, highly similar postures, for example, sitting and standing overlap significantly in the feature space and difficult to distinguish. Furthermore, there are some activities, such as walking upstairs, walking downstairs, walking in different postures, walking with carrying objects (backpack), walking while performing other activities (texting in smartphone), etc. have high motion similarities. These activities are challenging to separate as such activities share high similarity in the feature space because of their comparable movement patterns. In most of the cases, high correlations among activities are not uniform during the entire set of activities, which makes the recognition even harder. For example, sitting and standing actions are very much similar (difficult to distinguish), however, they are very different from walking (easily distinguishable).

Transitional activities can be further divided into four types [3]:

- Static to static postural transition (e.g., sit to stand, lying to sit, etc.)
- Static to dynamic (e.g., stand to walk)
- Dynamic to static (e.g., walk to stand)
- Dynamic to dynamic (e.g., walk to jog, jog to run, etc.)

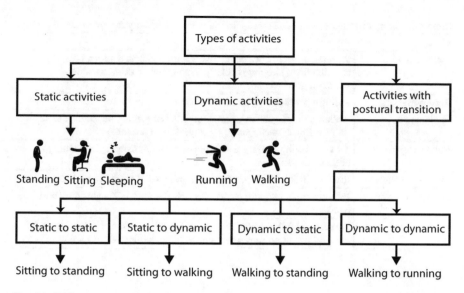

**Fig. 4.1**  Different types of daily human activities

**Fig. 4.2**  Primary
complication factors behind
sensor-based human activity
recognition (HAR)

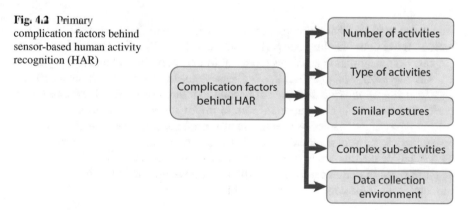

In many research projects, researchers have used the data collected from static to static postural transitions to distinguish static activities more efficiently. For example, to identify *sitting* and *lying* smartly, the past few time frames can be considered using the fact that for *sitting* activity, a previous postural transition (*standing* to *sitting*, or *lying* to *sitting*) is a required information. Similarly, to identify the dynamic activity (like, *walk*) well, we may consider the past few time frames using the fact that for *walking* activity, a previous *stand* is a must.

In Fig. 4.2, we have summarized the complication factors behind human activity recognition using sensor devices. These factors need to be considered with care to obtain better performance.

**Table 4.1**  Group of activities identified by sophisticated human activity recognition systems

| Group | Activities |
|---|---|
| Daily activities | Working at the PC, watching TV, watering plants, reading, brushing teeth, cleaning floor, sleeping, cooking, having lunch, drinking water, vacuuming. |
| Transportation | Riding a bus, subway, train, driving car, cycling, etc. |
| Ambulation | Walking, running, sitting, standing still, lying, climbing stairs, descending stairs, riding escalator, stay, jogging, riding elevator. |
| Phone usage | Text messaging, making a call, playing games, using apps, sending business e-mail. |
| Fitness/exercises | Rowing, lifting weights, spinning, nordic walking, doing push-up. |
| Nursing activities | Patient sitting, measure blood pressure, measure ECG, attaching bust bandage. |
| Falls | **Directional fall**: forward fall, backward fall, rightward fall, leftward fall, etc. **Fall based on movements**: stumbling, sliding, slipping, etc. |

## 4.2  Type of Activities

According to [9], the basic group of activities are shown in Table 4.1. Daily activities include basic activities that are performed each day inside the houses in most of the cases. Transportation activities represent human activities related to daily transports like *cycling*, *riding a bus*, *driving a car*, etc. In recent days, researchers are also focusing on *nursing activities* in hospitals or health care facilities, and *fitness activities* that can play an important role to improve our medical sectors. Besides several types of *fall detection* classes are also growing important attentions among researchers to prevent accidental fall of elderly people or patients in hospital. Collecting fall detection data is painful experience and difficult to do [10]. Among all of these types of activities, the recognition of ambulation activities like walk, stay, run, etc. is the most common research field.

### 4.2.1  Ambulation Activities

Ambulation activities can be classified into three basic types [3]:

- Static activities,
- Dynamic activities, and
- Activities with postural transitions as shown in Fig. 4.1.

Activities like sitting on a sofa, lying on a bed, standing, etc. are regarded as static activities where also postural transition can be presented (posture change from sit to stand or stand to sit). These activities can be classified easily from periodic activities, for example, walking, jogging, running, etc. because of the variation in terms of acceleration and velocity. However, extra complexities can be added in the classification procedure because of the presence of extremely comparable postures (standing and sitting). Furthermore, if there is a similarity in the feature space in the case of dynamic activities because of related action patterns, it also makes the classification procedure difficult.

In most of the cases, the similarity among performed activities is not uniform throughout the entire period and the entire set of activities. Due to this reason, the recognition of particular activities gets even harder. For instance, sitting and standing are much related (difficult to separate), however, they are very distinct from walking (easily separable).

## 4.3  Data Collection Protocol

Training data can be collected from several users either in the laboratory or free-living conditions. Laboratory data are obtained maintaining stringent protocols where activities are conducted at the same pace and for the same duration in constrained ways. But in the case of free-living conditions, subjects might act contrarily and in less restrained ways. Unsupervised, less-controlled and user-annotated data collection in case of long-term out-of-lab monitoring brings several challenges. Most crucial challenges include:

1. During out-of-lab monitoring, subjects tend to annotate data themselves without any supervision of researchers. This results in unreliable annotations creating difficulties for classifiers to be trained well, which eventually degrades the classifier's recognition accuracy.
2. Problems become more prominent as there are no standard ways to follow to perform a specific activity. For instance:

   - A person may *sit* on a sofa in such a way, which cannot be recognized to be either *lying* or *sitting*.
   - The same situation can occur in case of dynamic activities, such as *walking* can be recognized as *jogging* or *skipping* for some persons and vice versa.

Therefore, training and test dataset prepared in laboratory settings are easier to be recognized by machine learning algorithms than the dataset prepared in a free-living condition.

### 4.3.1  Basic Requirements

#### 4.3.1.1  Requirements of Training Data

We can evaluate recognition algorithms based on the variety and quantity of test data they demand. We can choose training data based on the following two criteria:

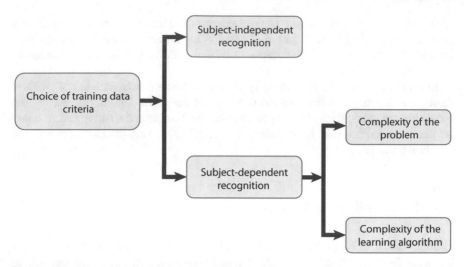

**Fig. 4.3** The basic criteria of training data collection

1. **Subject-independent recognition**: Any activity recognition algorithm is considered to be ideal if it is trained on a given subject population but it can identify activities for unseen subjects quite well. However, according to some previous works, such as [11], it is hard to achieve subject-independent recognition for a diverse set of activities as there are numerous ways of them to be performed by people. It requires a large number of training data and well-designed classifier to improve accuracy.

2. **Subject-dependent recognition**: Subject-dependent recognition is quite easier to achieve. Previous works on human activity recognition suggest that recognition algorithms function better with more person-specific training data. Ideally, the amount of data we require for better recognition depends both on the complexity of the problem and on the complexity of the chosen algorithm. These are:

   a. **The complexity of the problem**: In most of the cases, the nature and complexity of the problem is usually calculated by the unknown underlying feature which best relates the input variables to the output variable.

   b. **The complexity of the learning algorithm**: In most of the cases, the learning algorithm's complexity is calculated by the method used to inductively learn from concrete examples of the unknown underlying mapping function.

We have summarized the basic requirements of training data collection in Fig. 4.3, showing two criteria. Moreover, nonlinear algorithms, which are more powerful machine learning algorithms need more data as they are more flexible and non-parametric. They can find out how many parameters are needed to model a problem in addition to the values of those parameters. This added flexibility and power comes at the cost of requiring more training data.

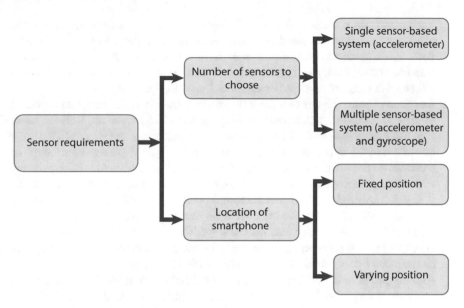

**Fig. 4.4** Factors that need to be considered to fulfill sensor requirement for HAR

### 4.3.1.2 Sensor Requirement

The recognition algorithm's sophistication will increase dramatically by the amount of smartphone sensors used, the form of sensors used, and the location of the device, when taking user data. We have summarized this situation in Fig. 4.4.

1. **The number of sensors to choose**: There are many sensors available for wearable systems and smartphones such as:

   - Global Positioning System (GPS),
   - Wi-Fi,
   - Bluetooth,
   - Accelerometers,
   - Magnetometers,
   - Gyroscopes,
   - Barometers,
   - Proximity sensors,
   - Temperature sensors,
   - Humidity sensors,
   - Ambient light sensors,
   - Cameras, and
   - Microphones.

   But we need to carefully select the amount of sensors because a recognition device employed with a limited range of sensors allows the operation in real-life implementation simpler and more efficient. Fewer sensor signals are needed to analyze a system with a small number of sensors than a large number of sensors. Although a system with fewer sensors is rendered easier by lower computing demand, the efficiency of identification of these systems is poorer than systems with a wide range of sensors because less knowledge is usable.

2. **Location of smartphone and wearable device**: Users usually carry smartphones in their shirt pocket, pant pocket, or keeping them in their hands. Besides, users who use wearable devices (like a smartwatch or other devices) wear it in the wrist or place it in the upper arm, leg, or ankle. Some people also keep smartphone or wearable devices in the waist. Accelerometers handle axis-based motion sensing, whereas, a gyroscope determines the orientation of the position of a smartphone. When collecting data, the orientation and position of sensing device should be noted, as data may vary for the device on various locations on the body of a person, even if the activity remains the same. Owing to the changing location of smartphones and other tracking devices, the overall quality of identification can be degraded by inaccurate detection of a specific incident.

Another problem is that certain people might not often carry their mobile with them when they're at home, rendering monitoring their actions difficult. In this situation, a wearable sensor may be a reasonable choice for certain people to wear it all day long when doing tasks, but it comes with the issue of discomfort. Especially, patients or elderly people find it difficult to use wearable devices all the time. In particular, people deny those wearable systems which hinder the daily physical behavior of subjects or force them into a fixed pattern of existence due to their size, methods of communication, or position in the body to wear that creates discomfort. These problems need to be considered with care by the fellow research community.

## 4.4   Design Issues

Main design issues concerning human activity recognition include the choice of sensors, the location of smartphones or wearable devices, the flexibility of wearable systems, the energy consumption of the device, etc. We have summarized these factors in Fig. 4.5.

### 4.4.1   Variety of Sensors

Triaxial accelerometers and gyroscopes are reasonably the most widely used sensors to recognize ambulation activities like walking, jogging, skipping, sitting, etc. These are usually less expensive, demand comparatively low power [12], and are installed in most of today's cellular phones and wearable devices. The Global Positioning System (GPS) facilitates all sort of location-based services and for context-aware

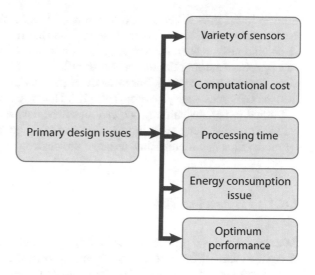

**Fig. 4.5** Factors that need to be considered in the case of human activity recognition research

purposes, including the recognition of the user's transportation mode [12]. The GPS sensor is very convenient. Tracking the place of users can be effective to understand their activities using ontological argumentation or reasoning [13]. For example, if a person is at a park, s/he is not brushing the teeth but might be moving or walking. Also, information about places can be obtained easily by using the Google Places Web Service [14], among other tools. Essential signs data (e.g., heart rate, respiration rate, skin temperature, skin conductivity, ECG, EEG, etc.) can also be monitored using sensors merged on a wearable device for medical activities recognition and monitoring patients [15].

### 4.4.2   Computational Cost

Computational cost is a key factor. Fewer sensor signals are needed to analyze a system with a small number of sensors. Though, lower computation requirement makes a system with fewer sensors simple, the recognition accuracy of such systems is inferior to the systems with a large set of sensors as less information is accessible.

### 4.4.3   Energy Consumption Issue

In case of human activity recognition, energy consumption is a vital issue to be considered as mobile devices such as sensors and cellular phones are generally energy-constrained. Short-range wireless networks (e.g., Bluetooth or Wi-Fi) should be fancied over long-range networks (e.g., cellular network or WiMAX) as the former requires lower power. Feature extraction and classification process can be done in the integration device as another approach so that raw signals would not have to be continuously sent to the server [13]. In the case of multiple sensors, there can be a system of turning the sensors off or reducing their sampling/transmission rate when they are not required. For instance, if the user's activity is sitting or standing still, the GPS sensor may be turned off [12]. Hence, by improving the design issues and utilizing nanotechnology, we can make the devices more energy-efficient. To reduce excessive energy consumption, we can take help of embedded programming for more efficient and long-time usage of the devices.

### 4.4.4  Processing Time

Processing time is a primary factor in the case of human activity recognition. Most of the use cases of this research area demand real-time performance without delay. If we focus only to improve the performance at the cost of higher processing time, this will fail to provide the result for the end-user on time. In the case of a medical emergency and accidental cases, this issue can be very serious. This is the reason to balance the trade-off between performance and processing time.

### 4.4.5  Optimum Performance

Human activity recognition application areas mostly cover patient-activity monitoring, fitness monitoring, accidental activity identification, etc. This is necessary to maintain optimum performance in real-time. False detection of activities (like human fall detection when the smartphone drops from a hand) and lower performance rate to detect activities can cause serious problem for patients. So, it is very important to ensure optimum performance while doing human activity recognition research. Some activities are for regular monitoring and some activities are for healthcare issues on serious cases. The purposes of the system will define the level of optimum performance of activity recognition.

## 4.5  Choosing the Position of the Smartphone

We have already discussed in the Introduction part about the possible positions (e.g., shirt pocket, left or right pants pocket, back pocket, holding in the left or right hand, carrying in a bag, etc.) of a smartphone to keep when a person is performing activities. The location of the device (smartphone or smartwatch) and, subsequently, the placing of the accelerometer or gyroscope is an important discussion point.One of the key points is that in order to quantify human acceleration, it is crucial to consider the movement of the human body and what physical properties one needs to evaluate [16]. This shows the significance of choosing the appropriate combination of measurement range and accelerometer placement.

Usually, body motion can be calculated very well with a single accelerometer positioned near the center of mass of the body situated within the pelvis [17]. The same theory goes for gyroscope too. Attachment to the waist facilitates the control of accelerations close to the center of mass, which is the benefit of this location. Every body motion can trigger a change in the center of mass [18]. Therefore, it is recommended that we try to keep the smartphone in a pant pocket up to waist height near the center of mass of the body to make the recognition more accurate. However, in reality, it is difficult to keep the sensors or mobile phones always in this position, especially for female (as in many countries, women wear different kinds of dresses that will not allow this setting).

There are other common positions for positioning, such as chest or leg [19]. Accelerometers usually ought to be connected to the portion of the body whose action is being observed. Ultimately, the best location to put the accelerometer must be decided depending on the requirement and the form of behaviors to be observed.

## 4.6   Conclusion

Due to the limitations and shortage of some comprehensive and large publicly available datasets on sensor-based human activity recognition, most of the researchers try to build their datasets. However, the collection of data in a proper way is a challenging task to deal with. In this chapter, we have analyzed those challenges and design issues along with some feasible solutions. Sensor requirement for data collection, the difference between lab-collected data and real-time data, sensor position, energy consumption issues, and other factors have been discussed in detail in this chapter.

## 4.7   Think Further

1. What factors do we need to consider while collecting human activity data?
2. What are the differences between data collected in a laboratory and free-living condition?
3. What is the relation between complexity and number of activities to classify?
4. What is the relation between complexity and types of activities?
5. Why it is difficult to classify activities with similar postures?
6. Why it is difficult to classify complex activities consisted of several sub-activities?
7. What are the different types of human activities?
8. What are the basic categories of ambulation activities?
9. What should be the ideal data collection protocol?

10. What are the basic requirements of training data?
11. What are the basic sensor requirements?
12. How number and types of sensors can bring a change in performance?
13. What are the primary design issues?
14. How can we handle computational cost in the field of HAR?
15. How we can deal with energy consumption issue in the field of HAR?
16. How to choose the ideal position of smartphone and wearable device for better performance?

# References

1. Antar, A.D., Ahad, M.A.R, Shahid, O.: Vision-based action understanding for assistive healthcare: a short review. In: IEEE CVPR Workshop (2019)
2. Ahad, M.A.R.: Vision and sensor based human activity recognition: challenges ahead (2020)
3. Antar, A.D., Ahmed, M., Ahad, M.A.R.: Challenges in sensor-based human activity recognition and a comparative analysis of benchmark datasets: a review. In: 2019 Joint 8th International Conference on Informatics, Electronics & Vision (ICIEV) and 2019 3rd International Conference on Imaging, Vision & Pattern Recognition (icIVPR), pages 134–139. IEEE, Cheney, WA (2019)
4. Ahad, M.A.R.: Motion History Images for Action Recognition and Understanding. Springer Science & Business Media, Berlin (2012)
5. Ahad, M.A.R.: Computer Vision and Action Recognition: A Guide for Image Processing and Computer Vision Community for Action Understanding, vol. 5. Springer Science & Business Media, Berlin (2011)
6. Hossain, T., Islam, M.S., Ahad, M.A.R., Inoue, S.: Human activity recognition using earable device. In: Proceedings of the 2019 ACM International Joint Conference on Pervasive and Ubiquitous Computing and Proceedings of the 2019 ACM International Symposium on Wearable Computers, pp. 81–84. ACM, New York (2019)
7. Tazin, T., Hossain, T., Ahad, M.A.R., Inoue, S.: Activity recognition by using lorawan sensor. In: 2018 ACM International Joint Conference on Pervasive and Ubiquitous Computing and 2018 International Symposium on Wearable Computers (UbiComp/ISWC) (2018)
8. Ahmed, M., Antar, A.D., Ahad, M.A.R.: An approach to classify human activities in real-time from smartphone sensor data. In: 2019 Joint 8th International Conference on Informatics, Electronics Vision (ICIEV) and 2019 3rd International Conference on Imaging, Vision Pattern Recognition (icIVPR), pp. 140–145 (2019)
9. Lara, O.D., Labrador, M.A.: A survey on human activity recognition using wearable sensors. IEEE Commun. Surv. Tutor. **15**(3), 1192–1209 (2013)
10. Islam, M.Z., Serikawa, S., Islam, Z.Z., Tazwar, S.M., Ahad, M.A.R.: Automatic fall detection system of unsupervised elderly people using smartphone. In: Annual Conference on Artificial Intelligence. IEEE, New York (2017)
11. Pantelopoulos, A., Bourbakis, N.G.: A survey on wearable sensor-based systems for health monitoring and prognosis. IEEE Trans. Syst. Man Cybernet. Part C (Appl. Rev.) **40**(1), 1–12 (2010)
12. Reddy, S., Mun, M., Burke, J., Estrin, D., Hansen, M., Srivastava, M.: Using mobile phones to determine transportation modes. ACM Trans. Sensor Netw. (TOSN) **6**(2), 13 (2010)
13. Riboni, D., Bettini, C.: Cosar: hybrid reasoning for context-aware activity recognition. Pers. Ubiquitous Comput. **15**(3), 271–289 (2011)
14. Google Places API.: http://code.google.com/apis/maps/documentation/places/. Accessed 17 March 2019

15. Yin, J., Yang, Q., Pan, J.J.: Sensor-based abnormal human-activity detection. IEEE Trans. Knowl. Data Eng. **20**(8), 1082–1090 (2008)
16. Park, S., Jayaraman, S.: Enhancing the quality of life through wearable technology. IEEE Eng. Med. Biol. Magaz. **22**(3), 41–48 (2003)
17. Sarela, A., Korhonen, I., Lotjonen, J., Sola, M., Myllymaki, M:. Ist vivago/spl reg/-an intelligent social and remote wellness monitoring system for the elderly. In: Information Technology Applications in Biomedicine, 2003. 4th International IEEE EMBS Special Topic Conference on, pp. 362–365. IEEE, Birmingham (2003)
18. Anguita, D., Ghio, A., Oneto, L., Parra, F.X.L., Ortiz, J.L.R.: Energy efficient smartphone-based activity recognition using fixed-point arithmetic. J. Univ. Comput. Sci. **19**(9), 1295–1314 (2013)
19. Vodjdani, N.: The ambient assisted living joint programme. In: Electronics System-Integration Technology Conference, 2008. ESTC 2008. 2nd, pp. 1–2. IEEE, London (2008)
20. Alain, C., et al.: Optimal cross-validation in density estimation with the $l\{2\}$ -loss. Annal. Stat. **42**(5), 1879–1910 (2014)

# Chapter 5
# Devices and Application Tools for Activity Recognition: Sensor Deployment and Primary Concerns

**Abstract** Sensory modality is a primary concern in sensor-based activity recognition research. The usage of wearable devices and utilizing embedded smartphone sensor data to recognize daily activities has become famous in this research field nowadays. This chapter deals with the challenges of choosing an appropriate sensing device and application tools for data collection. Sensing devices used in previous activity recognition research works have been described in detail with their hardware and software specification. Finally, the description and parameters of some important sensors (accelerometer, gyroscope, etc.) have been given.

## 5.1 Available Sensing Devices and Application Tools

Human Activity Recognition (HAR) has numerous important applications as well as, a number of challenges are ahead to deal with [1–8]. In this chapter, we present decides and application tools for human activity recognition—based on sensors. Most of the time tracking apps and devices combined on home appliances are used for the gathering of data from human behavior. But smartphones nowadays do provide the requisite sensors which can be used at any location at any time for tracking activities. But to access smartphone sensors, we need an application tool. We have discussed wearable devices and sensor modules in this segment, along with some applications and techniques that can be used for the identification of human activities. Researchers can use this sensing devices and apps for data collection or they can use the data collected by these devices for further research.

### 5.1.1 Node

Motion Node [9] is notably tiny in size (35 mm × 35 mm × 15 mm) and lightweight enough (14 g) to wear without difficulty for a prolonged period of time. This is a store-bought sensing platform to captivate human activity signals. Motion Node is a 6-DOF inertial measurement unit (IMU), which is specially intended for human

© Springer Nature Switzerland AG 2021                                                                 77
M. A. R. Ahad et al., *IoT Sensor-Based Activity Recognition*, Intelligent Systems
Reference Library 173, https://doi.org/10.1007/978-3-030-51379-5_5

motion sensing purposes. This device integrates a 3-axis accelerometer, a 3-axis gyroscope, and a 3-axis magnetometer. For each axis of accelerometer and gyroscope, the measurement range is ±6 g and ±500 dps respectively.

**Good Points:** Motion Node is a wired device and sends sampled sensor data to a laptop computer via a USB interface. No sensor data is missed in this case and the accuracy of the sensor data is well maintained.

**Challenge:** Possible anxiety is that the wire is cumbersome and may distort the sample data.

### 5.1.2   Eco Sensors

Eco sensors [10] are small (1 cm × 2 cm × 6 mm) in dimension and they have no internal battery. Their production price is relatively inexpensive. It consumes very little power. Each of the Eco sensors features a 2-axis accelerometer sensor equipped for the distinct task of tracking infant activity (Fig. 5.1).

**Good Points:** The compressed form factor and low energy consumption make Eco nodes suitable for several applications, e.g., environmental monitoring, medicine, ambient intelligence, and computer-human interface.

**Challenge:** One core concern is that the wireless range is limited to 10.7 m. Also, the multi-modal data collection is not allowed.

### 5.1.3   μParts

μParts [11] is a low-cost, compact sensor node (10 × 10) optimized for settings needing a high population of relatively low sampling rate sensors. μParts designers have restrained the components to a single side of the PCB, whereas putting the battery on the reverse side. These explicit design decisions have reduced the cost of the device to a great extent. To detect motion, this system includes temperature, light, and a ball switch sensor.

**Fig. 5.1**   Block diagram of Eco sensors

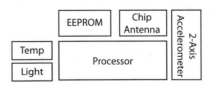

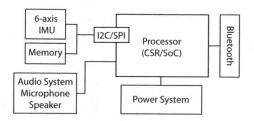

**Fig. 5.2** Schematic design of ESense

### 5.1.4 ESense

ESense [6, 12] is a wireless earbud for robust, effective, and multimodal sensing of human behavior. The purpose of this sensor is to track head and mouth-related action. This earbud can sense audio, orientation, change of movement, temperature, photoplethysmogram (PPG), and galvanic skin response. It uses Bluetooth communication for sending data to the receiver. This earbud can also be used for listening to music. The schematic design of ESense is shown in Fig. 5.2.

**Good Points:** ESense can be used as both for entertainment and sensing purposes. Also, the ear is relatively stationary than other body parts, so it will induce very little noise due to body jerking.

### 5.1.5 MITes (MIT Environmental Sensors)

It is a handheld package of omnipresent wireless sensing system equipped for the real-time compilation of human behaviors in natural settings for conduct research, built in MIT, USA [13]. It consists of 3.2 cm × 2.5 cm × 0.6 cm and 8.1 g (including battery) stick-on nodes that sense environmental or body information and transmit it wirelessly to one or several reception nodes. USB or RS232 serial ports are used to send the received data to the host computer. MITes have been developed using a common communication board with an easy-to-replace sensor connector, so several sensor nodes (light, temperature, etc.) can be accessed by removing just the onboard sensor and microcode.

### 5.1.6 MICA2DOT

MICA2DOT [14] is a wireless smart microsensor. It has a built-in temperature sensor and a battery monitoring system. It also contains 18 connecting pins, where different sensors can be attached to this sensor. This sensor also supports serial communication. There are three types of processor available for MICA2DOT (MPR500CA, MPR510CA, and MPR520CA), which mainly based on the Atmel ATmega 128L

**Fig. 5.3** Block diagram of
MICA2DOT

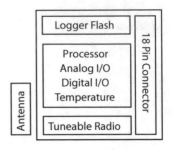

microcontroller. Diagram of MPR500CA is given in Fig. 5.3. A research team from
UC Berkeley developed an OS system for this sensor namely TinyOS. For trans-
mitting and receiving data this sensor can use Multi-channel Radio communication
using 868/916, 433 MHz or 315 MHz.
**Good Points:** MICA2DOT consumes a very little power. Also, it can communicate
with every node wireless as a router.
**Challenge:** The sensor has 6 analog input-output pins.

### 5.1.7 MICAz

MICAz [15] is an IEEE 802.15.4 measurement system with the architecture, particularly for intensely embedded sensor networks. It is very small in size and can
communicate wirelessly. There are 51 I/O pins, used as extension connectors for
Acceleration/Seismic, Magnetic, Light, Barometric Pressure, Temperature, Acoustic, RH and other Crossbow Sensor Boards. TinyOS operating system is used for
this sensor. The basic diagram of the MICAz is shown in Fig. 5.4. MICAz has I2C,
SPI and UART interfaces. The main application of this system is indoor building
monitoring and security.
**Good Point:** It can send or receive data at 250 kbps rate.

**Fig. 5.4** Block diagram of
MICAz

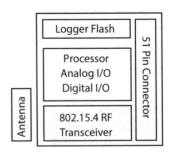

### 5.1.8  Intel Mote

Intel Mote [16] is a Bluetooth-based enhanced sensor network node. Intel Mote has increased CPU performance, improved radio bandwidth, and reliability along with a cheap cost. In this network node, there is an embedded wireless microcontroller consisting of an ARM7 core along with a Bluetooth radio, SRAM, and FLASH memory as well as different I/O choices. The design of this node also includes a reliable high bandwidth streaming transport layer. The software architecture used by the sensor is based on TinyOS.

**Good Points:** A new transportation protocol was built to enable end-to-end safe transfer of broad datagrams inside the network from one node to another arbitrary node. A low-power network configuration has been introduced, utilizing the Bluetooth hold mode. During this mode, data will still flow at extremely low speeds across the network, rising the resource usage while ensuring a quick response time for the network.

### 5.1.9  TMote Sky

TMote Sky [17] is an ultra-low-power IEEE 802.15.4 compliant wireless sensor module. It contains temperature, humidity, and light sensors with USB, which facilitates a broad range of mesh network applications. This module has out-of-the-box TinyOS support. TMote influences arising wireless protocols and the open-source software movement. It has USB support, optional SMA antenna, and 16-pin expansion support. The block diagram is shown in Fig. 5.5. In spite of featuring onboard sensors for robust application, cost and packaging size of TMote Sky is very little.

**Good Point:** It provides flexible interconnection with external sensors and peripherals.

### 5.1.10  Smart-ITS

Smart-ITS [18] is an integrated smart-object network. This represents the "Disappearing Computer" hypothesis, which places computation within the context of contact between people and their natural world. In the 1G Smart-ITS architecture, the three primary functionalities are mapped onto two hardware modules. Block diagram of a single module is shown in Fig. 5.6. One module is intended for communication purposes and the other for physical I/O. Both modules contain processors. An I2C data bus and a power bus interconnect the modules. The central module is called core board and it communicates with other Smart-ITS.

**Fig. 5.5** Block diagram of
TMote sky. PAR denotes
photosynthetically active
radiation sensor, and TSR is
total solar radiation sensor

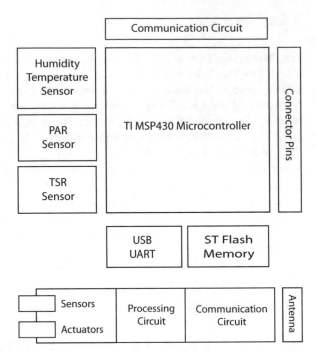

**Fig. 5.6** Block diagram of
Smart-ITS

### 5.1.11  Luna Nurse

Luna Nurse is a sensor that can monitor whether a person has left the bed or not
[19]. Three different states can be detected by this sensor by sensing the movement
of the person, which is: bed-rising up, sitting on the edge of the bed, and leaving the
bed. The detection method of this sensor is passive infrared. It can be installed in the
headrest side of the bed. The behavior on the bed can be detected within 2 m from
the sensor. The operating environment temperature should be 5 °C $\sim$ 40 °C and the
weight of this device is around 6 kg.

**Good Points:** It is easy to install the sensor from the bed and it can be easily set on
the head side of a person while sleeping. Being a non-contact type sensor without
sound, the user does not have to compromise comfort sleeping. This sensor can also
be used for fall accident prevention of children and elderly people while sleeping.
This product got a Good Design Award in 2014.

### 5.1.12  Google Glass App

A software has been developed at the University of Toronto, Canada for wearable
systems that help autistic children with social interactions. This [20] software is used
for tracking children who interact in real-world circumstances with adults. These

programs can be used for reinforcing strategies acquired in clinical environments in daily situations, such as home and education.

### 5.1.13 CC2650 SensorTag

The test package CC2650 SensorTag [21] acts as a sensor slave system built on a CC2650 multi-standard wireless MCU design based on Bluetooth low energy (BLE). It includes five peripheral sensors: IR temperature sensor, movement sensor, barometric pressure, humidity, and optical sensors. This package is designed with a robust sensor driver software solution, merged to a GATT server operating on TI BLE-Stack v2. For setup and data collection the GATT registry provides a rudimentary module for each sensor.

### 5.1.14 Device Analyzer

Device Analyzer [22] is a software that runs on Android smartphones running 2.1or higher and that gathers context consumption data when the handset is in operation. It has a personal analytics option that extracts data monitoring our everyday work like phone calls or our movement patterns using embedded phone sensors. Research works can be done using data collected by this application.

### 5.1.15 Smart Wearable Clothes

These clothes have brought a new era in the field of data collection related to activity monitoring [23]. These clothes use E-textile (electro-textile) and smart fabrics with electronics and other components embedded in. Smart clothes can monitor a user's fitness parameters during the workout. There are several types of smart wearable clothes, e.g., Nadi X Yoga Pants, Ambiotex, Owlet Smart Sock2, OMsignal Bra, Siren, Samsung NFC Suit, etc. Note that the *International Symposium on Wearable Computers* (ISWC) has been running for 24 years and this is the premier community for wearable technology-based research activities.

### 5.1.16 Maglietta Interattiva Computerizzata (MagIC)

MagIC [24] is an undershirt, which includes embedded sensors to measure heart rate and breathing rate. This product is made of conductive fibers. There is an electronic module consisting of a triaxial accelerometer, a data storage system, and a signal transmitter attached with the vest. The vest is waterproof (easy to wash) and comes in various sizes. The vest is specifically tailored to reduce artifacts. People with impaired movement can wear this cloth, as it has both sides and front opening with a zipper or velcro.

### 5.1.17 Cyber Glove II

This glove [25] is specifically designed to measure hand function of people with neurological disorders. Bend sensors were also used which contain an ink based on carbon/polymer whose resistance increases with bending. By using these bend sensors, this glove can measure finger and wrist flexion. The primary function is to track and monitor finger activity as the user completes different activities, so it can identify only minor differences in fine motor skills.

### 5.1.18 Bellabeat

Bellabeat [26] is a wellness-oriented outfit intended to build fashionable, compatible women's wearables. This sophisticated piece of jewellery is built to monitor health problems. This can be worn with accessories or objects, as a necklace or bracelet.

### 5.1.19 Fitness Trackers and Smartwatches

Samsung smartwatches (Gear Fit, Gear 2, and Gear 2 Neo), Apple Watch, Basis Peak fitness by Intel's new devices group, Fitbit, and Xiaomi Mi Bands are most popular for fitness tracking, daily movement monitoring, sleep monitoring, heart rate monitoring, etc. These devices can be used for data collection purposes [27].

### 5.1.20 HASC Tool and HASC Logger

HASC Tool [28] is an application built with Eclipse Plugin for the processing of action details. We can create a dataset on triaxial accelerometer data of different activities using this tool on a smartphone. HASC Logger, on the other hand, is a device that gathers data about operation via iPhone or iPod touch. Action label must be manually issued. It is explored by several works, e.g., [8].

## 5.1.21   LoRaWAN

LoRaWAN [29] is a module that can be interfaced with a number of sensors. Many research works have utilized this sensor for activity monitoring [7, 30]. It is a high-efficiency tool of optimum precision of calculation. LoRaWAN sensor has a network lifespan of up to 10 years, and the energy usage is much smaller compared to other sensor products. It provides real-time graphs of sensor data. This device has signal conditioning circuitry for analog signals and also has digital sensors interfacing capability.

## 5.1.22   Snapband

The [31] Snapband is a lightweight, multi-location contact input tool. This system can be snapped to various places very easily. Snapband addresses the challenges by utilizing current wearable gadgets that are limited to being carried at different places on the body (for example, wrist). This will restrict the capabilities for interaction based on physical limitations in body movement and positioning. On the other side, Snapband is a multi-functional wireless input system which can be worn at various locations of the body. It is also possible to mount this device onto objects in the environment. Users may change the position of the unit to various affordances, depending on specific conditions and usage cases.

## 5.1.23   LYRA

LYRA [32] is a smart wearable device, which has been designed for in-service flight assistant. This device assists flight attendants during their work. This device lets people use their smartphones to search and order services. Smart glasses and a smart shoe clip with RFID reader attachment give located details to flight attendants during long-distance flights.

## 5.1.24   YAWN

YAWN [33] is yet another wearable toolkit that simplifies the integration of micro-electronics into the fabric. YAWN is a bus-based, compact wearable toolkit that simplifies the interconnection by depending on a prefabricated fabric band constructed from three wires. This ensures fast reconfiguration, washability, and reduces the number of link issues.

## 5.2    Descriptions of Some Important Sensing Devices

In a signal acquisition system, sensors are the component of the instruments that are diligent in calculating variations in the spatial parameter and must be sensitive to the essence of the signal to be obtained [34]. The description of the sensors in this section addresses the challenges to correctly recognize static, dynamic and transitional activities performed by different subjects in different environments in the context of wearable sensors embedded in smart devices like smartphone and smartwatch. Some of these sensors can also be used as environmental sensors for monitoring human activities inside the home for patients and elderly people. Through using the Android sensor system, we can access certain sensors and acquire raw sensor data. The sensor architecture is part of the kit of android hardware and contains the following groups and interfaces: SensorManager, Sensor, SensorEvent, SensorEventListener, TriggerEvent, etc.

### 5.2.1    Accelerometer

Accelerometry is a movement kinematic analysis tool that enables the quantification of human body accelerations induced or sustained by the use of an accelerometer [35]. It has also been pointed out that, for long-term monitoring of human movement use of accelerometry is increasing rapidly [36]. The working principle is consistent with Newton's second law of motion. It measures the acceleration forces in $ms^{-2}$ that is applied to a device on all three physical axes (x, y, and z), as shown in Fig. 5.7, including the force of gravity, where $1\ g = 9.81\ ms^{-2}$. Accelerometers typically lead to good results in the detection of physical activity, which generally needs very little time, memory, which computing power [37]. The basic operational theory behind the accelerometer focused on Microelectromechanical Systems (MEMS) is the displacement of a tiny evidence mass carved into the silicone surface of the integrated circuit and supported by miniature beams [38].

**Fig. 5.7**  Triaxial accelerometer sensor

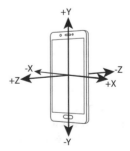

**Fig. 5.8**  Triaxial gyroscope
sensor

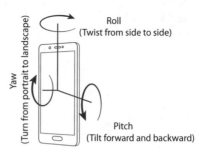

## 5.2.2  Gyroscope

A gyroscope identifies up/down, left/right and rotation around three axes (x, y, and z) as shown in Fig. 5.8 for more complex orientation details. It measures the influence of gravity applied to a system on all three physical axes (x, y, and z), in ms$^{-2}$. Inside phones gyroscopes do not use gears and gimbals much as conventional mechanical devices. Instead, they are gyroscopes of the Micro-Electro-Mechanical Systems (MEMS), a simplified variant of the principle mounted on an electronics board such that it may work within a smartwatch or mobile. This device can measure the angular velocity of the smartphone device, namely the pitch, roll, and yaw, which helps determine rapid shifts or small changes in the angular velocity of the user. When empirically analyzed, a periodic shift in gyroscopic data signals dynamic movement and small changes denote transition between static states of the user.

Early activity recognition algorithms relied solely on accelerometer data. However, the incorporation of gyroscope sensors in HAR adds another level of precision helping out the accelerometer with understanding which way the phone is orientated. It helps out in keeping the same recognition accuracy no matter what orientation the phone is kept by the user while taking data. As a result, use of both accelerometer and gyroscope is making a good impact on human activity recognition [39, 40].

## 5.2.3  Magnetometer

The tri-axis magnetometer utilizes the principle of Hall effect on a miniature scale to detect the effect of the earth's magnetic field along the three principal axes as shown in Fig. 5.9. The accelerometer, gyroscope, and magnetometer are housed in the same module using the MEMS technology.

**Fig. 5.9** Triaxial
magnetometer sensor

### 5.2.4 Pressure Sensor

Pressure sensor readings indicate the ambient air pressure near the device. Pressure sensor readings are useful in detecting the altitude of the device as shown in Fig. 5.10. It is essential to differentiate between the transportation modes of varying altitude levels, for example, buses and subway trains.

There are other types of sensor like Global Positioning System (GPS), Wi-Fi, Bluetooth, barometers, humidity sensors, light sensors, etc. in smart devices. However, these types of sensors are not widely used in wearable sensor-based human activity recognition domain but there can be possible research direction about utilizing data from these sensors for making a better activity recognition system. We can also extract data regarding linear acceleration, gravity, and orientation. However, these readings do not use a unique sensor, rather they are synthesized from the accelerometer, gyroscope and magnetometer data.

Linear acceleration provides us with the absolute acceleration of the device devoid of the gravity portion of acceleration data. Gravity readings specify the direction of the Earth's center. Orientation readings are derived from the gyroscopic data and present the azimuth angle of the device. The gravity and orientation sensors work in tandem to provide an accurate frame of reference for angular motions.

**Fig. 5.10** Pressure sensor
reading

## 5.3   Factors to Choose Accelerometer Sensor

Until picking a device, we have to weigh the following considerations [41] about the accelerometer sensor as seen in Fig. 5.11.

### 5.3.1   Frequency Response

Frequency Response is defined by the crystal density, piezoelectric properties, and case resonance frequency. It is the frequency range in which the accelerometer performance is within a defined variance, normally $\pm 5\%$. $g$ is an earth gravity acceleration of 32.2 ft/s$^2$, 386 in/s$^2$, *or* 9.8 m/s$^2$, respectively.

We need to consider the frequency range of the accelerometer to which it will detect motion and state a true output. Frequency response is measured in Hertz (Hz), which is normally specified as a range.

### 5.3.2   Dynamic Range

The maximal amplitude (can be positive or negative) which the accelerometer may calculate until distorting or clipping the output signal is called dynamic range. Normally dynamic range is defined in unit: g's.

### 5.3.3   Sensitive Axis

The inherent design of accelerometers tends it to detect inputs about an axis. We need to consider either the accelerometer is single-axis or triaxial. The most suitable accelerometers are triaxial accelerometers (TA) based on numerous applications, which can detect inputs in three orthogonal planes. Single-axis accelerometers, on the other side, will only track signals in one plane.

**Fig. 5.11** Important factors that need to be considered before choosing accelerometer sensor

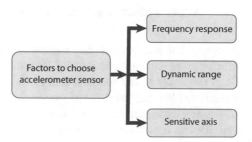

**Table 5.1** Target accelerometer system parameters

| Parameters | Target Value | | |
|---|---|---|---|
| Number of axes | 3 | | |
| Sampling frequency | 1.25–800 Hz | | |
| | | 12 bits | 8 bits |
| Maximum acceleration amplitude and | ±2 g | 2/2048 | 2/128 |
| Acceleration resolution (in bits and g) | ±4 g | 4/2048 | 4/128 |
| | ±8 g | 8/2048 | 8/128 |
| Maximum acceleration without damage | 5000 g | | |

Most of the accelerometers used in day-to-day activity tracking calculate either vertical (uni-axial) or tri-axial (triaxial) accelerations and respond to movement frequency and speed [42]. But it needs to be considered that, as most of the human motion occurs in more than one movement axis, triaxial accelerometers are used to measure the acceleration in each orthogonal axis. Most of the smartphone's accelerometers are the triaxial accelerometer. Adapted from [41], Table 5.1 provides several characteristics of the accelerometer used in different investigations.

The reference point is usually chosen in such a way that 0 g correlates to a freefall state and the final output number represents the highest sum of g the system can register. An accelerometer's efficiency depends on four factors: the direction in which it is positioned (decided by the smartphone's size), its orientation at this spot, the subject's posture, and the operation being carried out by the subject [43]. The accelerometer efficiency, when the subject is at rest, is calculated by its inclination relative to the gravitational vector. If we know the accelerometer's orientation relative to the human, then we may use the resulting accelerometer recordings to evaluate the subject's posture relative to each position [43].

## 5.4   Acceleration Related to Human Activities

The magnitude of the acceleration decreases from the head to the feet and is typically greater in the vertical direction [44]. To the other side, frequency continues to decline from the ankle to the shoulder and is higher in the vertical direction than in the transverse [44]. The precision of recognition activity as a feature of the sampling rate of the accelerometer was examined by [45]. It has been shown that for ambulation activities no significant improvement in accuracy is obtained above 20 Hz. Adapted

**Table 5.2** Amplitude and movement while performing physical activities for different sensor locations

| Motion | Vertical (g) | | | Horizontal (g) | | |
|---|---|---|---|---|---|---|
| | Head | Body | Ankel | Head | Body | Ankel |
| Walking | – | –0.3; 0.8 | –1.7; 3.3 | –0.2; 0.2 | –0.3; 0.4 | –2.1; 0.4 |
| Running | 0.8; 4.0 | 0.9; 5.0 | 3.0; 12.0 | – | – | – |

from [44], Table 5.2 indicates certain amplitude cycles for certain standard activities; all units are in $g$.

The measuring ranges influence the precision and expense of the accelerometer, and the frequency defines the device's sampling rate which is critical during the sensor design process. Accelerometers ought to be able to measure accelerations of $\pm 12\,g$, in general, and higher than $\pm 6\,g$, if put at the waist to test daily life events [46].

## 5.5 Conclusion

Due to the advancements of sensor-based technologies and smart devices, researchers are focusing on creating sensor-based activity recognition dataset. In this chapter, we have provided different types of sensor modules and mobile application tools that can be utilized by the researchers for collecting data. This chapter also provides a brief description of important sensors like accelerometer, gyroscope, magnetometer, pressure sensor, etc. However, the devices and systems may have newer versions as time passes. Moreover, some systems may be unavailable in the market in the future. Therefore, these systems and tools are depicting the recent and current trends, from which we can understand an overall view of the devices and tools for human activity recognition arena.

## 5.6 Think Further

1. How to choose sensor modalities for human activity recognition?
2. What are the available sensing devices and application tools?
3. Which mobile applications can be used to collect smartphone sensor data?
4. Which software can be used to collect data from wearable and environmental sensors?
5. Describe some important sensing devices.
6. What is the purpose of using acceleration data in HAR?
7. What is the purpose of using gyroscope data in HAR?
8. What is the purpose of using magnetometer data in HAR?

9. What is the purpose of using pressure sensor data in HAR?
10. What factors need to be considered to choose sensor for better performance?
11. What are the limitations or constraints of using accelerometer in HAR?
12. What are the limitations or constraints of using gyroscope in HAR?
13. What are the limitations or constraints of using magnetometer in HAR?
14. What are the limitations or constraints of using pressure sensor in HAR?
15. What are the recent smart sensors for HAR?
16. What are the recent tools and mobile applications for HAR?
17. Write down various sensors that are suitable for various activity types. Explain the reasons for your selections of sensors.
18. Are there any sensors that can be explored for HAR (but not yet exploited by the research community)?

# References

1. Antar, A.D., Ahad, M.A.R., Shahid, O.: Vision-based action understanding for assistive health-care: a short review. In: IEEE CVPR Workshop (2019)
2. Ahad, M.A.R.: Vision and sensor based human activity recognition: challenges ahead (2020)
3. Antar, A.D., Ahmed, M., Ahad, M.A.R.: Challenges in sensor-based human activity recognition and a comparative analysis of benchmark datasets: a review. In: 2019 Joint 8th International Conference on Informatics, Electronics and Vision (ICIEV) and 2019 3rd International Conference on Imaging, Vision and Pattern Recognition (icIVPR), pp. 134–139. IEEE (2019)
4. Ahad, M.A.R.: Motion History Images for Action Recognition and Understanding. Springer Science & Business Media, Berlin (2012)
5. Ahad, M.A.R.: Computer Vision and Action Recognition: a Guide for Image Processing and Computer Vision Community for Action Understanding, vol. 5. Springer Science & Business Media, Berlin (2011)
6. Hossain, T., Islam, M.S., Ahad, M.A.R., Inoue, S.: Human activity recognition using ear-able device. In: Proceedings of the 2019 ACM International Joint Conference on Pervasive and Ubiquitous Computing and Proceedings of the 2019 ACM International Symposium on Wearable Computers, pp. 81–84. ACM (2019)
7. Tazin, T., Hossain, T., Ahad, M.A.R., Inoue S.: Activity recognition by using lorawan sensor. In: 2018 ACM International Joint Conference on Pervasive and Ubiquitous Computing and the 2018 International Symposium on Wearable Computers (UbiComp/ISWC) (2018)
8. Ahmed, M., Antar, A.D., Ahad, M.A.R.: An approach to classify human activities in real-time from smartphone sensor data. In: 2019 Joint 8th International Conference on Informatics, Electronics Vision (ICIEV) and 2019 3rd International Conference on Imaging, Vision Pattern Recognition (icIVPR), pp. 140–145 (2019)
9. Motionnode IMU platform. http://www.motionnode.com/. Accessed 20 Mar 2019
10. Park, C., Liu, J., Chou, P.H.: Eco: an ultra-compact low-power wireless sensor node for real-time motion monitoring. In: Proceedings of the 4th International Symposium on Information Processing in Sensor Networks, p. 54. IEEE Press (2005)
11. Beigl, M., Decker, C., Krohn, A., Riedel, T., Zimmer, T.: $\mu$parts: low cost sensor networks at scale. In: Ubicomp 2005. Citeseer (2005)
12. Kawsar, F., Min, C., Mathur, A., Montanari, A.: Earables for personal-scale behavior analytics. IEEE Pervas. Comput. **17**(3), 83–89 (2018)

13. Tapia, EM., Marmasse, N., Intille, S.S., Larson, K.: Mites: Wireless portable sensors for studying behavior. In: Proceedings of Extended Abstracts Ubicomp 2004: Ubiquitous Computing (2004)
14. Mica2dot wireless microsensor mote. https://www.willow.co.uk/mpr5x0-_mica2dot_series.php, (2005). Accessed 22 Mar 2019
15. Micaz wireless measurement system. http://www.cmt-gmbh.de/Produkte/WirelessSensorNetworks/MPR2400.html (2005). Accessed 22 Mar 2019
16. Kling, R.M. et al.: Intel mote: an enhanced sensor network node. In: International Workshop on Advanced Sensors, Structural Health Monitoring, and Smart Structures, pp. 12–17 (2003)
17. Moteiv. tmote sky: Ultra low power IEEE 802.15.4 compliant wireless sensor module. http://www.moteiv.com/products/docs/tmote-skydatasheet.pdf (2005). Accessed 22 Mar 2019
18. Beigl, M., Gellersen, H.: Smart-its: An embedded platform for smart objects. In: Smart Objects Conference (sOc), vol. 2003 (2003)
19. Luna nurse. http://www.g-mark.org/award/describe/41326?locale=en. Accessed 22 Mar 2019
20. Google glass app. http://glass-apps.org/google-glass-application-list. Accessed 22 Mar 2019
21. Cc2650 sensortag. http://processors.wiki.ti.com/index.php/CC2650_SensorTag_User's_Guide. Accessed 23 Mar 2019
22. Device analyzer. http://deviceanalyzer.cl.cam.ac.uk/. Accessed 23 Mar 2019
23. Smart wearable clothes. https://www.wareable.com/smart-clothing/best-smart-clothing. Accessed 23 Mar 2019
24. Magic (maglietta interattiva computerizzata). http://www.ncbi.nlm.nih.gov/pubmed/20421189. Accessed 25 Mar 2019
25. Cyberglove 2. http://www.cyberglovesystems.com/products/cybergl Accessed 25 Mar 2019
26. Bellabeat. http://gadgetsandwearables.com/bellabeat/. Accessed 25 Mar 2019
27. Fitness trackers and smart watches. https://www.cnet.com/topics/wearable-tech/buying-guide/. Accessed 25 Mar 2019
28. Hasc tool and hasc logger. http://hasc.jp/tools/hasctool-en.html. Accessed 25 Mar 2019
29. The lorawan sensor. https://www.decentlab.com/lorawan/. Accessed 02 Mar 2019
30. Ahad, M.A.R., Hossain, T., Tazin, T., Inoue, S.: Study of lorawan technology for activity recognition. In: 2018 ACM International Joint Conference on Pervasive and Ubiquitous Computing and the 2018 International Symposium on Wearable Computers (UbiComp/ISWC) (2018)
31. Dobbelstein, D., Arnold, T., Rukzio, E.: Snapband: a flexible multi-location touch input band. In: Proceedings of the 2018 ACM International Symposium on Wearable Computers, pp. 214–215. ACM (2018)
32. Auda, J., Hoppe, M., Amiraslanov, O., Zhou, B., Knierim, P., Schneegass, S., Schmidt, A., Lukowicz, P.: Lyra: smart wearable in-flight service assistant. In: Iswc'18: Proceedings of the 2018 Acm International Symposium on Wearable Computers. Association for Computing Machinery, pp. 212–213 (2018)
33. Thar, J., Stönner, S., Heller, F., Borchers, J.: Yawn: yet another wearable toolkit. In: Proceedings of the 2018 ACM International Symposium on Wearable Computers, pp. 232–233. ACM (2018)
34. Lane, N.D., Miluzzo, E., Lu, H., Peebles, D. Choudhury, T., Campbell, A.T.: A survey of mobile phone sensing. IEEE Commun. Mag. **48**, 140–150
35. Kavanagh, J.J., Menz,H.B.: A technique for quantifying movement patterns during walking. In: Gait and Posture, pp. 1–15 (2008)
36. Zheng, Y., Wong, W.-K., Guan, X., Trost, S.: Physical activity recognition from accelerometer data using a multi-scale ensemble method. In: Twenty-Fifth Annual Conference on Innovative Applications of Artificial Intelligence. IAAI (2013)
37. Mannini, A., Intille, S.S., Rosenberger, M., Sabatini, A.M., Haskell, W.: Activity recognition using a single accelerometer placed at the wrist or ankle. In: Medicine and Science in Sports and Exercise (2013)
38. Wu, J., Gang, P., Daqing, Z., Guande, Q., Shijian, L.: Gesture Recognition with a 3-d Accelerometer. Ubiquitous Intelligence and Computing, pp. 25–38. Springer, Berlin (2009)
39. Reyes-Ortiz, J.-L., Oneto, L., Ghio, A., Sama, A., Anguita, D., Parra, X.: Human activity recognition on smartphones with awareness of basic activities and postural transitions. In: International Conference on Artificial Neural Networks, pp. 177–184 (2014)

40. Bahle, G., Gruenerbl, A., Lukowicz, P., Bignotti, E., Zeni, M., Giunchiglia, F.: Recognizing hospital care activities with a coat pocket worn smartphone. In: 6th International Conference on Mobile Computing, Applications and Services (MobiCASE), IEEE, pp. 175–181 (2014)
41. Shotton, J., Fitzgibbon, A., Cook, M., Sharp, T., Finocchio, M., Moore, R., Kipman, A., Blake, A.: Real-time human pose recognition in parts from single depth images. In: 2011 IEEE Conference on Computer Vision and Pattern Recognition (CVPR), pp. 1297–1304. IEEE (2011)
42. Tapia, E.M., Intille, S.S., Larson, K.: Activity recognition in the home using simple and ubiquitous sensors. In: International Conference on Pervasive Computing, pp. 158–175. Springer, Berlin (2004)
43. Najafi, B., Aminian, K., Paraschiv-Ionescu, A., Loew, F., Bula, C.J., Robert, P.: Ambulatory system for human motion analysis using a kinematic sensor: monitoring of daily physical activity in the elderly. IEEE Trans. Biomed. Eng. **50**(6), 711–723 (2003)
44. The house of the future. http://architecture.mit.edu/house_n. Accessed 02 Mar 2019
45. The gator-tech smart house. http://www.icta.ufl.edu/gt.htm. Accessed 02 Mar 2019
46. The inhaus project in germany. http://www.inhaus.fraunhofer.de/. Accessed 02 Mar 2019

# Chapter 6
# Sensor-Based Benchmark Datasets: Comparison and Analysis

**Abstract** Human Activity Recognition (HAR) using installed sensors has made renowned progress in the field of pattern recognition and human-computer interaction. To make efficient machine learning models, researchers need publicly available benchmark datasets. In this chapter, we have bestowed a comprehensive survey on sensor-based benchmark datasets. We have not considered RGB or RGB-Depth video-based action or activity-related datasets in this book. We have performed a complete analysis of benchmark datasets, that incorporates information about sensors, attributes, activity classes, etc. These datasets sum up a good number of sensor-based daily activities, medical activities, fitness activities, device usage, fall detection, transportation activity, and hand gesture data.

## 6.1 Benchmark Datasets Information

Human Activity Recognition (HAR) has numerous important applications as well as, a number of challenges are ahead to deal with [1–8]. In this chapter, we present sensor-based benchmark datasets and their comparative presentations. Researchers face multiple obstacles including technological difficulties, the question of anonymity, corporate authority, etc. in the data collection process [3]. However, with these challenges, a large range of databases are currently accessible for sensor dependent behavior recognition. We evaluated benchmark datasets from well-known databases in this segment, with related details such as data set characteristics, attribute characteristics, number of attributes, classes, and so on. Moreover, we have also mentioned the devices and sensors information used for collecting data in these datasets. These datasets contain data related to daily activities, fall classification, nursing and medical activities, sports activities, physical activities, sleep monitoring systems, and so on. Smartphones, wearable devices, and multiple environmental sensor-based smart home devices have been used mainly for collecting data in these datasets.

© Springer Nature Switzerland AG 2021      95
M. A. R. Ahad et al., *IoT Sensor-Based Activity Recognition*, Intelligent Systems
Reference Library 173, https://doi.org/10.1007/978-3-030-51379-5_6

## 6.2   UCI Machine-Learning Repository

University of California Irvine (UCI) machine learning repository [9] contains datasets in different domains, usually in *.CSV format. Datasets in this collection provide details regarding categories of attributes, missing values, target area, etc. The flexibility of data description and application richness are the key advantages in these datasets. We have listed datasets from this repository in Tables 6.1 and 6.2 for smartphone and wearable sensor-based movement identification. These datasets are publicly available. The names of these listed datasets are

- Heterogeneity Human Activity Recognition (HHAR),
- Activity recognition with healthy older people using a Battery-less Wearable Sensor (UCIBWS),
- Activity Recognition system based on Multisensor data fusion (AReM),
- Human Activity Recognition using smartphones (HAR),
- Smartphone-based recognition of Human Activities and Postural Transitions (HAPT),
- Activity recognition from Single Chest-mounted accelerometer (Single Chest),
- OPPORTUNITY activity recognition (OPPORTUNITY),
- Activities of Daily Living (ADLs) recognition using binary sensors,

**Table 6.1** List of publicly available 15 UCI ML repository datasets with basic information

| Dataset | Subjects | Activities | Attributes | Instances | Year |
|---|---|---|---|---|---|
| HHAR [19] | 9 | 6 | 16 | 43930257 | 2015 |
| UCIBWS [20] | 14 | 4 | 9 | 75128 | 2016 |
| AReM [21] | 1 | 6 | 6 | 42240 | 2016 |
| HAR [22] | 30 | 6 | 561 | 10299 | 2012 |
| HAPT [23] | 30 | 12 | 561 | 10929 | 2015 |
| Single Chest [24] | 15 | 07 | N.A | N.A | 2014 |
| OPPORTUNITY [13] | 4 | 35 | 242 | 2551 | 2012 |
| ADLs [25] | 2 | 10 | N.A | 2747 | 2013 |
| REALDISP [26] | 17 | 33 | 120 | 1419 | 2014 |
| UIFWA [24] | 22 | 2 | N.A | N.A | 2014 |
| PAMAP2 [27] | 9 | 19 | 52 | 3850505 | 2012 |
| DSA [28] | 8 | 19 | 5625 | 9120 | 2013 |
| Wrist ADL [28] | 16 | 14 | 3 | N.A | 2014 |
| RSS [29] | N.A | 2 | 4 | 13197 | 2016 |
| MHEALTH [30] | 10 | 12 | 23 | 120 | 2014 |
| WISDM [31] | 51 | 18 | 6 | 15630426 | 2019 |
| WESAD [32] | 15 | 3 | 12 | 63000000 | 2018 |
| SFDLA [33] | 17 | 36 | 138 | 3060 | 2018 |

N.A: Not available

**Table 6.2**  List of publicly available 15 UCI ML repository datasets with sensor information

| Dataset | Devices | Sensors/module | Missing values |
| --- | --- | --- | --- |
| HHAR [19] | 4 smartwatch and 8 smartphone | Accelerometer and gyroscope | Yes |
| UCIBWS [20] | Battery-less wearable sensor | 7 RFID reader antennas at two rooms | No |
| AReM [21] | Chipcon AT86RF230 radio subsystem with a TinyOS firmware | WSN (RSS data collection using IRIS nodes) at 20 Hz | No |
| HAR [22] | Waist-mounted smartphone (Samsung Galaxy S II) | Accelerometer and gyroscope at 50 Hz | No |
| HAPT [23] | Waist-mounted smartphone (Samsung Galaxy S II) | Accelerometer and gyroscope at 50 Hz | No |
| Single chest [24] | Chest-mounted device | Accelerometer 52 Hz | No |
| Opportunity [13] | Wearable object | Accelerometers, motion sensors, ambient sensors, etc | Yes |
| ADLs [25] | WSN device | 12 sensors (PIR, magnetic, pressure and electric sensors) | No |
| REAL-DISP [26] | Wearable device | 9 sensors (accelerometers, 4D quaternions, magnetic, gyroscope, etc.) | No |
| UIFWA [24] | Android smartphone at chest pocket | Accelerometer | No |
| PAMAP2 [27] | 3 inertial measurement units and a heart rate monitor | Wireless IMU and heart rate monitoring sensors | Yes |
| DSA [28] | Five Xsens MTx units | 9 sensors per unit. (accelerometers, gyroscopes, and magnetometers) at 25 Hz | No |
| Wrist ADL [34] | Wrist-worn device | Accelerometer | No |
| RSS [29] | WSN device in office environment | 5 sensors | No |
| MHEALTH [30] | Wearable device Shimmer2 (BUR10) | Accelerometer, 2 lead ECG, and magnetic sensors | No |
| WISDM [31] | Smartphone and smartwatch | Accelerometer, gyroscope | No |
| WESAD [32] | Wrist and chest worn device | RespiBAN, and Empatica E4 | Yes |
| SFDLA [33] | Sensors fixed in 6 positions | 23 sensing features | Yes |

- REALDISP activity recognition (REALDISP),
- User Identification From Walking Activity (UIFWA),
- PAMAP2 Physical Activity Monitoring (PAMAP2),
- Daily and Sports Activities (DSA),
- Dataset for ADL recognition with Wrist-worn accelerometer (Wrist ADL),
- Indoor user movement prediction from RSS data (RSS),
- MHEALTH dataset (MHEALTH).
- WISDM smartphone and smartwatch activity biometrics (WISDM),
- Wearable Stress and Affect Detection (WESAD), and
- Simulated Falls and Daily Living Activities (SFDLA).

Among these datasets, HHAR, OPPORTUNITY, PAMAP2, and SFDLA datasets contain missing values. There are many ways to handle missing values like list-wise or case deletion, pairwise deletion, maximum likelihood, and so on according to the research work [10–12]. This is an active area of research in sensor-based human activity recognition where the training dataset contains missing values [13–18].

## 6.3   Pervasive System Research Group

Pervasive System Research Group[1] has several datasets, source codes, and apps, which with a proper assertion, can be utilized for research purposes. This is a research group of the University of Twente, The Netherlands. In Tables 6.3 and 6.4, we have summarized the activity datasets from this research group. These datasets are publicly available. The listed datasets are

- Smoking Activity,
- Complex Human Activities,
- Physical Activity Recognition, and
- Sensors Activity datasets.

## 6.4   Human Activity Sensing Consortium (HASC)

HASC[2] is a non-profit voluntary organization, which conducts activities to build large-scale databases using wearable sensors for human behavior monitoring. We have listed datasets from HASC in Tables 6.5 and 6.6. These datasets are

- HASC2010corpus,
- HASC2011corpus,
- HASC2012corpus,

---

[1]http://ps.ewi.utwente.nl/Datasets.php.
[2]http://hasc.jp/.

**Table 6.3** List of publicly available Pervasive system research group datasets with basic information

| Dataset | Subjects | Activities | Applications | Year |
|---|---|---|---|---|
| Smoking activity [35] | 11 | 5 | Detection of smoking activity | 2016 |
| Complex human activities [36] | 10 male age: 23–35 | 13 | Complex human activities recognition | 2016 |
| Physical activity recognition [37] | 4 | 6 | Daily human activities recognition | 2013 |
| Sensors activity [38] | 10 age: 25–30 | 7 | Healthcare | 2014 |

**Table 6.4** List of publicly available pervasive system research group datasets with sensor information

| Dataset | Devices | Sensors/module |
|---|---|---|
| Smoking activity [35] | Smartwatch at wrist position and smartphone in pocket | Accelerometer, gyroscope, linear acceleration sensor, and magnetometer |
| Complex human activities [35] | Two mobile phones (Samsung Galaxy S2) in right pocket and wrist | Accelerometer, gyroscope, linear acceleration sensor, magnetometer |
| Physical activity recognition [37] | 4 smartphones on four body positions (wrist, jeans pocket, arm, belt) | Accelerometer, gyroscope, magnetometer at 50 Hz sampling frequency |
| Sensors activity [38] | 5 smartphones on five body positions | Accelerometer and gyroscope |

- HASC-IPSC: Indoor Pedestrian Sensing Corpus,
- HASC-PAC2016: Pedestrian Activity Corpus, and
- HASC-BDD: Ballroom Dance Dataset.

Among these datasets, HASC2010corpus, HASC-IPSC, and HASC-BDD are publicly available and explored by researchers, e.g., [8, 10].

## 6.5 Medical Activities-Related Datasets

Recognition of clinical events has various implications for continuous control of heart attack patients, pregnant women, and the elderly. We have also mentioned several datasets related to medical practices in Tables 6.7 and 6.8 namely,

**Table 6.5** List of HASC datasets with basic information

| Dataset | Subjects | Activities | Applications | Year | Public availability |
|---|---|---|---|---|---|
| HASC 2010 corpus [39] | 540 | 6 | Human activity recognition using smartphone or terminal devices | 2010 | Yes |
| HASC 2011 corpus [40] | 116 | 6 | Same as HASC 2010 corpus | 2011 | No |
| HASC 2012 corpus [41] | 136 | 6 | Same as HASC 2010 corpus | 2012 | No |
| HASC IPSC [42] | 107 | 6 | Indoor position and building structure estimation research | 2014 | Yes |
| HASC PAC [43] | 510 | 6 | Human activity recognition related to Pedestrian | 2016 | No |
| HASC BDD [44] | 7 | 13 | Dance step recognition in ballroom | 2019 | Yes |

**Table 6.6** List of HASC datasets with sensor information

| Dataset | Devices | Sensors/module |
|---|---|---|
| HASC 2010, 2011, 2012 corpus [39–41] | Smartphone (pocket, bag) and HASC tool | Triaxial accelerometer. Sampling rate: 10–100 Hz |
| HASC IPSC [42] | Two mobile phones (back waist pocket, bag) and HASC logger | Accelerometer, Wi-Fi, pressure sensor, angular velocity, and geomagnetism sensor |
| HASC PAC [43] | Smartphone (waist, arm, bag, chest, foot) and HASC logger | Accelerometer, gyro, magnetometer, sensor (pressure, proximity, light), and Wi-Fi |
| HASC BDD [44] | Wearable sensors and video camera | Six inertial sensors |

- Daphnet Freezing of Gait (FoG) dataset,
- Nursing Activity dataset,
- Nursing Care Records [45], and
- Predicting Parkinson's Disease dataset.

Among these datasets, Daphnet Freezing of Gait Data Set (Daphnet FoG) and Predicting Parkinson's Disease datasets are publicly available.

**Table 6.7**  List of medical activities-related datasets with basic information

| Dataset | Subjects | Activities | Applications | Year |
|---|---|---|---|---|
| Daphnet FoG [46] | 10 Parkinson's disease (PD) patients | 3 | Monitoring PD patients' walk and detecting sporadic freezing of gait | 2010 |
| Nursing activity [47] | Labelled data (22 nurses), unlabelled data (60 nurses) | 25 | Monitoring nursing activities in the hospital | 2016 |
| Predicting Parkinson's disease [48] | 9 Parkinson's disease (PD) patients | 2 | Monitoring and measure symptoms of PD disease | 2011–12 |

**Table 6.8**  List of medical activities-related datasets with sensor information

| Dataset | Devices | Sensors/module |
|---|---|---|
| Daphnet FoG [46] | Wearable device that can obtain gait data | Accelerometer sensors. At foot, knee, and hip |
| Nursing activity [47] | Mobile devices (IPod) in breast pockets and accelerometers on right wrist and back waist | Accelerometer sensor for nurses combined with wearable and environmental sensors for patients |
| Predicting Parkinson's disease [48] | Smartphones for at least one charge cycle per day (about 4–6 h) | Accelerometer, compass, ambient light, proximity, battery level, GPS, and audio sensors |

## 6.6   Physical and Sports Activities-Related Datasets

Physical and sports activities related to datasets can be used for the research work of monitoring daily work-out time to remain healthy and fit. Many wearable devices are available nowadays, which come with the feature of fitness tracking. But, most of the cases, these devices fail to distinguish among similar complex physical activities. So, research works in this area have many opportunities. In Tables 6.9 and 6.10, we have summarized 4 datasets related to physical activities monitoring. All datasets are publicly available. The listed datasets are,

- Body Attack Fitness dataset,
- Swimmaster dataset,
- Fitbit dataset, and
- BWT (Body Weight Training) dataset.

**Table 6.9** List of publicly available fitness activities-related datasets with basic information

| Dataset | Subjects | Activities | Applications | Year |
|---|---|---|---|---|
| Body attack fitness [49] | 1 | 6 | Leg-based physical activities monitoring and correct sensor placement research | 2009 |
| Swimmaster [50] | 12 (5 × 25 m) crawl | Swim activities | Evaluate swim techniques, specially the body balance and body rotation during crawl swimming | 2009 |
| Fitbit [51] | 30 Fitbit users | 11 | Tracking different. Fitbit trackers from the variation of outputs and behaviour tracking of people | 2016 |
| BWT [52] | N.A | 10 | Support system for beginners to perform effective Body Weight Training (BWT) alone | 2018 |

**Table 6.10** List of publicly available fitness activities-related datasets with sensor information

| Dataset | Devices | Sensors/module |
|---|---|---|
| Body attack fitness [49] | Wearable system for legs | 10 accelerometer sensors on the leg |
| Swimmaster [50] | Body attached device | 5 accelerometer sensors at the right and left wrist, at the lower and upper back, and one at the head |
| Fitbit [51] | Fitbit (activity tracking wrist band) | Accelerometer, gyroscope, sleep monitoring sensor, heart rate monitor, etc |
| BWT [52] | SenStick sensor device | Accelerometer, gyroscope, magnetic, temperature, humidity, pressure, light, and UV sensors |

## 6.7 Household Activities-Related Datasets

With the great advancement in the field of Internet of Things (IoT), the smart home concept with multiple sensors embedded in home appliances, vehicles, etc. are getting more famous. These sensors' data can be used for tracking daily household activities, which will suggest the users a well-suited daily routine for the rest of the week. We have listed some datasets based on environmental sensors and sensors merged in home appliances in Tables 6.11 and 6.12. Among these datasets, all are publicly available except Activity Recognition with Ambient Sensing (ARAS) dataset [53]. The names of the enlisted datasets are,

- MIT PlaceLab dataset,
- CMU Multi-Modal Activity Database (CMU-MMAC) dataset,
- Ubicomp dataset,
- Activity Recognition with Ambient Sensing (ARAS) dataset,
- Smart Home dataset,
- Center for Advanced Studies in Adaptive Systems (CASAS) dataset,

**Table 6.11** List of household activities-related datasets with basic information

| Dataset | Subjects | Activities | Applications | Year |
|---|---|---|---|---|
| MIT PlaceLab [54] | 1 | 10 | Household activities recognition | 2006 |
| CMU-MMAC [55] | 43 | 5 | Cooking activities recognition | 2009 |
| Ubicomp [56] | 3 | 10, 13, and 16 activities in 3 houses | Monitoring daily activities of inhabitants inside their houses | 2008 |
| Smart home [57] | 2 age: 26 and 57, in 2 houses | 17 | Sensor based activity recognition in bathroom and kitchen | 2010 |
| ARAS [53] | Multiple from 2 houses | 27 | Human activity recognition from ambient sensors merged at home | 2013 |
| CASAS [58] adults | Younger and older | 11 | Common activities detection that span multiple environment settings | 2012 |
| Home sensors [59] | 2 women in two homes | 16 | Proactive care for the aging based on environment sensors | 2004 |
| Daily routine [60] | 1 | 34 | Recognize daily routines as a probabilistic combination of activity patterns | 2008 |

**Table 6.12** List of 8 household activities-related datasets with sensor information

| Dataset | Devices | Sensors/module |
|---|---|---|
| MIT PlaceLab [54] | Wearable system | 5 accelerometers (one on each limb and one on the hip) and one wireless heart rate monitor |
| CMU-MMAC [55] | Multi-sensor network based devices | Accelerometers, gyroscopes, video, audio, RFID tags, motion capture system based on body markers, skin temperature and galvanic skin response (GSR) |
| Ubicomp [56] | RFM DM 1810 WSN kits with a rich API and blue-tooth headset (Jabra BT250v) for annotation | House A: 14 sensors, House B: 23 sensors, House A: 21 sensors |
| Smart home [57] | WSN with many nodes, which communicates with a central gateway | Sensors attached in 2 bathrooms and 2 kitchens. Reed switches, mercury contacts, PIR, and float sensors |
| ARAS [53] | Home sensors based system, that communicates using ZigBee protocol wirelessly | 20 binary sensors for each house. Force sensor, photocell, contact sensors, proximity sensors, temperature sensors, sonar sensors, infrared receivers, etc |
| CASAS [58] | Sensor network in house and data stored in SQL | 475 sensors used for creating 11 datasets in this research |
| Home sensors [59] | Sensor device and EMS tool | Home 1: 77 environmental sensors, Home 2: 84 environmental sensors (installed on doors, windows, oven, cabinets, drawers, etc.) |
| Daily routine [60] | Porcupine sensor platform with PIC microcontroller | 3D accelerometer, temperature sensor, and light sensor |

- Cooking Activity dataset[3]
- Home Sensors dataset, and
- Daily Routine dataset.

## 6.8  Device Usage Activities-Related Datasets

Recently, many users use devices like smartwatches, smartphones, tabs, etc. Data related to the activities of users while using these smart devices can be a promising field of research. With user-agreement to avoid privacy issues, these research works can help the users in contextual life management, choosing apps and the best phone

---

[3]Cooking Activity Challenge with *International Conference on Activity and Behavior Computing* (ABC), 2020 https://abc-research.github.io/cook2020.

**Table 6.13**  List of device usage activities-related datasets with basic information

| Dataset | Subjects | Activities | Applications | Year |
|---|---|---|---|---|
| Activity classification [61] | 10 | 7 | Determine user activities while wearing a smartwatch or using a phone | 2015 |
| Reality mining [62] | 100 | 5 | Sensing complex social systems and social patterns in daily user activity | 2004 |
| MDC [63] | 200 | 7 | Contextual life management, health, and well-being | 2009–2010 |
| Device analyzer [64] | 31320 | 4 | Phone plan and app recommendation | 2014 |
| Insight for wear [65] | 11.5 million records | 7 | Help users augmenting their own behavior or device use | 2015 |

plans suited for them based on historical use. In Tables 6.13 and 6.14, we have listed these kinds of datasets related to user activity while using smart devices with embedded sensors. The listed datasets are,

**Table 6.14**  List of device usage activities-related datasets with sensor information

| Dataset | Devices | Sensors/module | Availability |
|---|---|---|---|
| Activity classification [61] | Smartphone in 5 body locations | Accelerometer. Sampling frequency: 50 Hz | Publicly available |
| Reality mining [62] | 100 Nokia 6600 phones with the Symbian OS and IBM laptop | Proximity, GPS, call log, SMS, etc | Not publicly available |
| MDC [63] | Nokia phones | Sensors embedded in Nokia phone | Only for non-profit organizations |
| Device analyzer [64] | Smartphone and app. | GSM, GPS | Yes (with condition) |
| Insight for wear [65] | Lifelogging smartwatch app. | Embedded sensors in smartwatch | Yes (with condition) |

- Activity Classification dataset,
- Reality Mining dataset,
- Mobile Data Challenge (MDC) dataset,
- Device Analyzer dataset, and
- Insight for Wear dataset.

## 6.9    Wearable Sensor-Based Datasets

Wearable devices have to be built taking into account consumer versatility. In several databases, low weight, trendy, and convenient wearable tools with integrated sensors were used for behavior tracking. Some of these publicly accessible databases have been addressed with specific details in Tables 6.15 and 6.16. The names of such datasets mentioned are,

- University of Southern California Human Activity Dataset (USC-HAD),
- UC Berkeley WARD (Wearable Action Recognition Database),
- Skoda mini checkpoint dataset,
- Human-Computer Interaction gestures dataset (HCI),
- PPS Grouping dataset,
- University of Dhaka—Mobility Dataset (DU-MD),
- Human Gait Database (HuGaDB), and
- UTD Multimodal Human Action Dataset (UTD-MHAD).

## 6.10    Smartphone Sensor-Based Datasets

Smartphones have been most people's regular companion during the day. So, using embedded sensor data from smartphones, tracking user behaviors is convenient. But, there are some problems as the precise and accurate classification of activities depends on the placement or location of the smartphone (for example, shirt pocket, carrying in hand or bag, thigh pocket, or elsewhere) and orientation of the smartphone. Moreover, we have to rely on embedded sensors only. Yet mobile sensor-based behavior detection has clutched other research interests because of the flexibility and cost-effective method. We also mentioned several publicly accessible datasets of this method in Tables 6.17 and 6.18. There are some more datasets based on smartphone sensor, e.g.,

- Datasets in UCI ML repository [9],
- Human Activity Sensing Consortium, e.g., HASC2010 Corpus [39],
- HASC-IPSC (Indoor Pedestrian Sensing Corpus) [42]

Some of these datasets can be presented under different categories due to similar modalities or patterns. Some other datasets are,

**Table 6.15**  List of publicly available wearable sensor-based datasets with basic information

| Dataset | Subjects | Activities | Applications | Year |
|---|---|---|---|---|
| USC-HAD [66] | 14 | 12 | Healthcare (physical fitness monitoring and elder care) | 2012 |
| WARD [67] | 20 | 13 | Human action recognition | 2009 |
| Skoda [68] | 1 | 10 | Activities related to car maintenance | 2007 |
| HCI [49] | 1 | 5 | Human-computer interaction based on hand gesture | 2009 |
| PPS grouping [69] | 10 | 2 | Detection of different group formation while walking | 2009 |
| DU-MD [70, 71] | 25 | 10 | Used for training HAR package in existing fitness bands to act as remote healthcare monitoring system | 2018 |
| HuGaDB [72] | 18 | 12 | Health-care-related studies, such as in walking rehabilitation or Parkinson's disease recognition | 2017 |
| UTD-MHAD [73] | 8 (4 male, 4 female) | 27 | Robust human action recognition using fusion of data from differing modality sensors | 2015 |

- WISDM (Wireless Sensor Data Mining) dataset,
- Smart Devices dataset,
- Gait database.

Among these datasets, the Gait database by Institute of Scientific and Industrial Research (ISIR), Osaka University is the largest database with 745 subjects (388 males and 357 females) [74, 75].

## 6.11  Locomotion and Transportation Datasets

In recent years, there have been various researches on multimodal sensor data that are collected during transportation and locomotion activities. Transportation mode recognition is beneficial in sectors regarding travels and tourism, traffic condition monitoring, user transportation behavior monitoring, collecting data for transport

**Table 6.16**  List of publicly available wearable sensor-based datasets with sensor information

| Dataset | Devices | Sensors/module |
|---------|---------|----------------|
| USC-HAD [66] | Motion node (high performance inertial sensing device) | Triaxial accelerometer and gyroscope |
| WARD [67] | DexterNet (sensor platform) | Triaxial accelerometer and biaxial gyroscope |
| Skoda [68] | Wearable system | 20 triaxial accelerometer on both right and left arm |
| HCI [49] | Wearable system for hand | 8 triaxial accelerometer on both arms |
| PPS grouping [69] | Wearable system | 10 triaxial accelerometer sensor on hip |
| DU-MD [70] | Trillion node engine project's devices from The University of Tokyo | Accelerometer |
| HuGaDB [72] | Body sensor network. | 3 pairs of inertial sensors (accelerometer, gyroscope), and 1 pair of EMG sensors |
| UTD-MHAD [73] | Kinect camera and wearable device | RGB, depth, skeleton from Kinect; 9-axis MEMS sensor (3-axis acceleration, 3-axis angular velocity, and 3-axis magnetic strength) |

**Table 6.17**  List of 3 publicly available smartphone sensor-based ambulation activity datasets with basic information

| Dataset | Subjects | Activities | Applications | Year |
|---------|----------|------------|--------------|------|
| WISDM [76] | 29 | 6 | Smartphone sensor-based activity monitoring | 2012 |
| Smart devices [77] | N.A | 11 | Human behaviour recognition using smart devices | 2017 |
| Gait database [78] | 745 (388 males and 357 females) | 5 gait activities | Analyzing the dependence of gait authentication performance | 2013 |

**Table 6.18** List of publicly available smartphone sensor-based ambulation activity datasets with sensor information

| Dataset | Devices | Sensors/module |
|---------|---------|----------------|
| WISDM [76] | Several android phones, like the Nexus One, HTC Hero and Motorola Backflip | Triaxial accelerometer |
| Smart devices [77] | Smartphone, smartwatch and smartglass along with 2 apps (SWIPE and Timelogger) | Accelerometer, gyroscope, Electrooculography (EOG), and pressure sensor |
| Gait database [78] | Motorola ME860 smartphone | 3 Inertial sensors (Accelerometer and gyroscope) and a smartphone (accelerometer) at 100 Hz |

statistics, etc. *Smart City* is a big term recently and these datasets are very crucial for the future development of the Smart City concept. We can also easily access sensor data to monitor the average speed of traffic in a region. Public transit providers may enhance reliability by changing their routes and schedules, based on locomotion and transportation details. There are also many context-aware applications including,

- Activity and health monitoring [79],
- Parking spot detection [80], and
- Content delivery optimization [81, 82].

The available datasets are very limited for the general analysis of locomotion and usage of transportation modes (e.g., public transport, bike, car, etc.). Tables 6.19 and 6.20 list some locomotion and transportation datasets. Researchers explored the SHL dataset for locomotion and transportation study [83–86]. The listed datasets in these tables are done by Yang, Wang, Reddy, Siirtola, Hemminki, Zhang, Xia, Widhalm, Jahangiri, Su, Geolife, HTC, US Transportation, and SHL (Sussex-Huawei Locomotion-Transportation) dataset [83, 84, 84, 86].

## 6.12 Datasets on Fall Detection Techniques

Accidental fall detection of patients, pregnant women, and elderly people have become a major focus of attention among researchers due to its life-saving applications [100]. In this regard, access to public databases can establish an extensive and systematic assessment of fall detection techniques. In Tables 6.21 and 6.22, we have summarized publicly available datasets [101] on numerous types of falls along with similar activities closely related to fall.

However, it is extremely difficult to prepare fall detection datasets—as you need the subjects to fall down in different directions and various ways [100]. These can be varied, and painful for the subjects to do. The fall by a healthy person, and the

**Table 6.19** List of the locomotion and transportation datasets with basic information

| Dataset | Subjects | Activities | Labelled data | Public | Year |
|---|---|---|---|---|---|
| By Yang [87] | 3 | 6 | 3 | No | 2009 |
| By Geolife [88] | 182 | 6 | 50176 | Yes | 2010 |
| By Wang [89] | 7 | 6 | 12 | No | 2010 |
| By Reddy [90] | 16 | 5 | 120 | No | 2010 |
| By Siirtola [91] | 5 | 6 | 3 | No | 2012 |
| By Hemminki [92] | 16 | 6 | 150 | No | 2013 |
| By Zhang [93] | 15 | 6 | 30 | No | 2013 |
| By Xia [94] | 18 | 4 | 22 | No | 2014 |
| HTC [95] | 224 | 10 | 8311 | No | 2014 |
| By Widhalm [96] | 15 | 8 | 355 | No | 2014 |
| By Jahangiri [97] | 10 | 5 | 25 | No | 2015 |
| By Su [98] | 5 | 6 | 3 | No | 2016 |
| US transportation [99] | 13 | 5 | 31 | Yes | 2018 |
| SHL [83] | 3 | 28 | 2812 | Yes | 2018–2019 |

**Table 6.20** List of locomotion and transportation datasets with sensor and device information

| Dataset | Devices | Sensors/module |
|---|---|---|
| By Yang [87] | 1 smartphone, no preferred placement | Accelerometer |
| By Geolife [88] | 1 GPS logger or 1 GPS phone | GPS |
| By Wang [89] | 1 smartphone, no preferred placement | Accelerometer |
| By Reddy [90] | 6 smartphones in 6 positions: arm, waist, chest, hand, pocket, and bag | Accelerometer, GPS, Wi-Fi, and GSM |
| By Siirtola [91] | 1 smartphone in trousers | Accelerometer |
| By Hemminki [92] | 1 smartphone with partially no preferred placement and partly fixed locations: trousers, bag, torso | Accelerometer, GPS |
| By Zhang [93] | 1 smartphone, no preferred placement | Accelerometer |
| By Xia [94] | 1 smartphone, jacket/torso | Accelerometer, gyroscope, and GPS |
| HTC [95] | 1 smartphone, no fixed position | Accelerometer, gyroscope, and magnetometer |
| By Widhalm [96] | 1 smartphone, no fixed position | Accelerometer, GPS |
| By Jahangiri [97] | 1 smartphone, no preferred placement | Accelerometer, gyroscope, and GPS |
| By Su [98] | 1 smartphone, no preferred placement | Accelerometer, gyroscope, magnetometer, and barometer |
| US transportation [99] | 1 smartphone, no fixed position | 9 smartphone sensors |
| SHL [83] | 4 smartphones in 4 positions: hand, torso, backpack, and trousers. 1 front facing camera | 15 smartphone sensors, 1 time lapse video |

**Table 6.21** List of publicly available datasets for fall detection techniques with basic information

| Dataset | Subjects | Activities/falls | Applications | Year |
| --- | --- | --- | --- | --- |
| DLR [102] | 19 | 15 activities, 1 falls | Activity monitoring along with fall detection | 2010 |
| MobiFall [103] | 24 | 9 activities, 4 falls | Fall detection using smartphones | 2013 |
| UR fall [104] | 6 males | 5 activities, 4 falls | Falls monitoring in office and home environment | 2014 |
| Tfall [105] | 10 | 8 falls | Use of compensation strategies to prevent the fall | 2014 |
| TST fall [106] | 11 | 4 activities, 4 falls | Fall activity monitoring | 2014 |
| Cogent labs [107] | 42 | 15 activities, 4 falls | Wearable sensor-based fall monitoring for elder people | 2015 |
| Gravity [108] | 2 | 7 activities, 12 falls | Numerous types of fall monitoring | 2015 |
| DMPSBFD [109] | 5 martial artists | 10 activities, 4 falls | Daily activities and fall monitoring | 2015 |
| UMA fall [110] | 17 | 8 activities, 3 falls | Falls and daily activities monitoring | 2016 |
| MobiAct [111] | 57 | 9 activities, 4 falls | Fall detection using smartphones | 2016 |
| UniMiB SHAR [109] | 30 | 9 activities, 8 falls | Preventing fall accidents | 2017 |
| Sis fall [112] | 38 | 19 activities, 15 falls | Different types of fall monitoring | 2017 |

fall by an elderly or a patient are not the same at all in terms of speed, patterns of falling down, etc. And we can not create a more 'realistic dataset' with elderly or sick people for fall detection! It is not possible. One alternative way is to exploit mannequins in a controlled manner, or to develop any robotic system that can be managed with some realistic features like weight, walking speed, patterns of fall down, etc. Computer-generated simulated data may be explored as well to see the results (similar to some synthetic video-based action recognition datasets).

## 6.13 UCR Time Series Classification Archive

This is a repository hosted by the University of California, Riverside that represents time-series data, which comes in two parts namely train and test set [113]. The two files are in the same format but are generally of different sizes. The files are in the

**Table 6.22** List of publicly available datasets for fall detection techniques with sensor information

| Dataset | Devices | Sensors/module |
|---|---|---|
| DLR [102] | 1 external IMU (Xsens MTx) | Accelerometer, gyro, and magnetometer at 100 Hz |
| MobiFall [103] | 1 smartphone (Samsung Galaxy S3) | Accelerometer (87 Hz) and gyroscope (100 Hz) |
| UR fall [104] | 1 external IMU (x-io-x-IMU) (Bristol, UK) | Accelerometer. Aampling frequency: 256 Hz |
| Tfall [105] | 1 smartphone (Samsung Galaxy Mini) | Accelerometer. Sampling frequency: 45 Hz |
| TST fall [106] | 2 external IMU, (Shimmer device) | Accelerometer. Sampling frequency: 100 Hz |
| Cogent labs [107] | 2 SHIMMER sensor nodes with intelligence modularity for health sensing | Triaxial accelerometer, triaxial gyroscope, and bluetooth device. Sampling frequency: 100 Hz |
| Gravity [108] | Samsung Galaxy S3 and LG G Watch | Accelerometer. Sampling frequency: 50 Hz. |
| DMPSBFD [109] | 5 smartphones. | Accelerometer and gyroscope. Sampling frequency: 5 Hz |
| UMA fall [110] | 1 Smartphone (Samsung Galaxy S5) 4 external IMUs (LG G4, Bosch accelerometer, Texas Instruments SensorTag, and InveSense MPU-9250) | Accelerometers, gyroscope, and magnetometer. Sampling frequency: 100 Hz for smartphone and 20 Hz for IMU |
| MobiAct [111] | 1 smartphone (Samsung Galaxy S3) | Accelerometer (87 Hz) and gyroscope (100 Hz) |
| UniMiB SHAR [109] | 1 smartphone (Samsung Galaxy Nexus) | Triaxial accelerometer. Sampling frequency: 50 Hz |
| Sis fall [112] | 1 external sensing mote | 2 accelerometers and 1 gyroscope. Sampling frequency: 200 Hz |

standard ASCII format that can be read directly by most tools/languages. UCR[4] repository contains electrocardiogram (ECG) data, robots' surface detection data, etc. that are measured using different devices.

## 6.14   PhysioBank

This domain focuses on smartphones and wearable devices data for monitoring health and wellness improvement. PhysioBank[5] is a rich source of psychological health-related signals along with numerous benchmark datasets. There are datasets includ-

---

[4]http://www.cs.ucr.edu/~eamonn/time_series_data/.
[5]https://www.physionet.org/physiobank.

ing electrocardiogram (ECG), electroencephalogram (EEG), gait data of human in different stages, and other forms of medical data. Most of the medical data are of multivariate time series characteristics [114].

## 6.15  CRAWDAD

Most of the data in the Community Resource for Archiving Wireless Data At Dartmouth (CRAWDAD)[6] domain is related to mobile devices and wireless networks. There is no limitation on the data format. Sometimes, this domain provides data about location or mobility of users who use mobile devices with embedded sensors like accelerometer, GPS, gyroscope, etc. [115].

## 6.16  Conclusion

In this chapter, we have presented a comprehensive survey on sensor-based benchmark datasets, covering about more than 150 datasets. Many papers explored their datasets but these are usually neither robust nor open for others to explore. Because of the challenges to find publicly available benchmark datasets on sensor-based activity recognition, most of the researchers face difficulties in this field of research. In this chapter, we have summarized the basic information of numerous benchmark sensor-based datasets to recognize human activities. Beside that, we have provided the source of these datasets with proper analysis so that the researchers can easily select an appropriate dataset according to their research goals.

These datasets sum up several types of sensor-based daily activities, medical activities, fitness activities, device usage, fall detection, and hand gesture data. Therefore, these datasets remain only to the authors or to their groups and not open for all to exploit. Some industries have introduced some datasets through various on-body sensor devices, however, these are for their internal research purposes as well and not open for all. Most of the datasets are smaller in size in terms of data, number of classes, number of subjects, variabilities, etc. Therefore, it is necessary to make collaborations for having full or part of those industrial large dataset to start research activities. Various competitions and challenges are explored in different conferences or workshops based on some datasets (e.g., SHL challenge with ACM UbiComp, Cooking Activity Challenge[7]). These are more helpful for research communities. Apart from the above points, we have not explored any RGB or RGB-Depth video-based action- or activity-related datasets in this chapter or book. Some of these works are available in various papers or books [1, 2, 4, 5]. There are demands for datasets in

---

[6]http://crawdad.org/.

[7]Cooking Activity Challenge with *International Conference on Activity and Behavior Computing* (ABC), 2020 https://abc-research.github.io/cook2020.

the domain of various falls in hospital and outside, activities related to rehabilitation progress in rehabilitation centers, autism study, eye movement and tracking to understand behavior [116–118], earable-based datasets for challenging activities that are usually not possible to manage by other sensors [6, 119], fatigue and tiredness-related datasets [120, 121], and so on.

## 6.17   Think Further

1. Why do we need the information of benchmark datasets in the field of sensor-based human activity recognition?
2. How can we classify sensor-based activity datasets?
3. What should be the key factors to choose a dataset?
4. Which information are needed to compare different datasets?
5. Mention some application areas of UCI Machine-Learning repository datasets.
6. Mention some application areas of Pervasive System Research Group datasets.
7. Mention some application areas of Human Activity Sensing Consortium (HASC) datasets.
8. Mention some application areas of medical activities-related Datasets
9. Mention some application areas of physical and sports activities-related datasets.
10. Mention some application areas of household activities-related datasets.
11. Mention some application areas of device usage activities-related datasets.
12. Mention some application areas of wearable sensor-based datasets.
13. Mention some application areas of smartphone sensor-based datasets.
14. Mention some application areas of locomotion and transportation-related datasets.
15. Mention some application areas of fall detection datasets.
16. If you want to make a new dataset for sensor-based activity recognition, then what kind of datasets do you prefer to develop? Explain the goal or purpose for each target dataset.
17. List up the problems of the existing datasets and make a guideline for future datasets to develop.
18. What are the unrealistic points in the existing dataset?
19. How can you develop a dataset having more than thousand subjects?
20. How can you develop a dataset for fall detection purpose?
21. How can you develop a dataset combining vision-based systems (e.g., RGB camera or depth camera) and sensor-based system?
22. What are the missing points in the vision-based as well as sensor-based activity datasets?
23. Mention some applications where both vision-based and sensor-based action classes can be explored?
24. Which activities are more important than other activity classes in the existing datasets?
25. Enlist 30 activity classes that are explored already along with the datasets.

26. Enlist top 10 activity classes that will be more important in the near-future and that are more realistic.

# References

1. Antar, A.D., Ahad, M.A.R., Shahid, O.: Vision-based action understanding for assistive healthcare: a short review. IEEE CVPR Workshop (2019)
2. Ahad, M.A.R.: Vision and sensor based human activity recognition: challenges ahead (2020)
3. Antar, A.D., Ahmed, M., Ahad, M.A.R.: Challenges in sensor-based human activity recognition and a comparative analysis of benchmark datasets: a review. In: 2019 Joint 8th International Conference on Informatics, Electronics and Vision (ICIEV) and 2019 3rd International Conference on Imaging, Vision and Pattern Recognition (icIVPR), pp. 134–139. IEEE (2019)
4. Ahad, M.A.R.: Motion History Images for Action Recognition and Understanding. Springer Science & Business Media, Berlin (2012)
5. Ahad, M.A.R.: Computer Vision and Action Recognition: a Guide for Image Processing and Computer Vision Community for Action Understanding, vol. 5. Springer Science & Business Media, Berlin (2011)
6. Hossain, T., Islam, M.S., Ahad, M.A.R., Inoue, S.: Human activity recognition using earable device. In: Proceedings of the 2019 ACM International Joint Conference on Pervasive and Ubiquitous Computing and Proceedings of the 2019 ACM International Symposium on Wearable Computers, pp. 81–84. ACM (2019)
7. Tazin, T., Hossain, T., Ahad, M.A.R., Inoue, S.: Activity recognition by using lorawan sensor. In: 2018 ACM International Joint Conference on Pervasive and Ubiquitous Computing and the 2018 International Symposium on Wearable Computers (UbiComp/ISWC) (2018)
8. Ahmed, M., Antar, A.D., Ahad, M.A.R.: An approach to classify human activities in real-time from smartphone sensor data. In: 2019 Joint 8th International Conference on Informatics, Electronics Vision (ICIEV) and 2019 3rd International Conference on Imaging, Vision Pattern Recognition (icIVPR), pp. 140–145 (2019)
9. Lichman. Uci machine learning repository. http://archive.ics.uci.edu/ml, 2013. Accessed 25 Mar 2019
10. Hossain, T., Goto, H., Ahad, M.A.R., Inoue, S.: A study on sensor-based activity recognition having missing data. In: 2018 Joint 7th International Conference on Informatics, Electronics and Vision (ICIEV) and 2018 2nd International Conference on Imaging, Vision and Pattern Recognition (icIVPR), pp. 556–561. IEEE (2018)
11. Ahad, M.A.R., Hossain, T., Tazin, T., Inoue, S.: Study of lorawan technology for activity recognition. In: 2018 ACM International Joint Conference on Pervasive and Ubiquitous Computing and the 2018 International Symposium on Wearable Computers (UbiComp/ISWC) (2018)
12. Savvaki, S., Tsagkatakis, G., Panousopoulou, A., Tsakalides, P.: Matrix and tensor completion on a human activity recognition framework. IEEE J. Biomed. Health Inf. 21(6), 1554–1561 (2017)
13. Chavarriaga, R., Sagha, H., Calatroni, A., Digumarti, S.T., Tröster, G., del R Millán, J., Roggen, D.: The opportunity challenge: a benchmark database for on-body sensor-based activity recognition. Pattern Recognit. Lett. 34(15), 2033–2042 (2013)
14. Akhavian, R., Behzadan, A.: Wearable sensor-based activity recognition for data-driven simulation of construction workers' activities. In: 2015 Winter Simulation Conference (WSC), pp. 3333–3344. IEEE (2015)
15. Yin, J., Yang, Q., Pan, J.J.: Sensor-based abnormal human-activity detection. IEEE Trans. Knowl. Data Eng. 20(8), 1082–1090 (2008)
16. Wang, L., Gu, T., Tao, X., Lu, J.: Sensor-based human activity recognition in a multi-user scenario. In: European Conference on Ambient Intelligence, pp. 78–87. Springer (2009)

17. Pham, C., Diep, N.N., Phuong, T.M.: A wearable sensor based approach to real-time fall detection and fine-grained activity recognition. J. Mob. Multimedia **9**(1&2), 15–26 (2013)
18. Tao, G., Wang, L., Zhanqing, W., Tao, X., Jian, L.: A pattern mining approach to sensor-based human activity recognition. IEEE Trans. Knowl. Data Eng. **23**(9), 1359–1372 (2010)
19. Blunck, H., Bhattacharya, S., Stisen, A., Prentow, T.S., Kjærgaard, M.B., Dey, A., Jensen, M.M., Sonne, T.: Activity recognition on smart devices: dealing with diversity in the wild. GetMobile: Mob. Comput. Commun. **20**(1), 34–38 (2016)
20. Torres, R.L.S., Ranasinghe, D.C., Shi, Q., Sample, A.P.: Sensor enabled wearable RFID technology for mitigating the risk of falls near beds. In: 2013 IEEE International Conference on RFID (RFID), pp. 191–198. IEEE (2013)
21. Palumbo, F., Gallicchio, C., Pucci, R., Micheli, A.: Human activity recognition using multisensor data fusion based on reservoir computing. J. Ambient Intell. Smart Environ. **8**(2), 87–107 (2016)
22. Anguita, D., Ghio, A., Oneto, L., Parra, X., Reyes-Ortiz, J.L.: A public domain dataset for human activity recognition using smartphones. In: ESANN (2013)
23. Reyes-Ortiz, J.-L., Oneto, L., Samà, A., Parra, X., Anguita, D.: Transition-aware human activity recognition using smartphones. Neurocomputing **171**, 754–767 (2016)
24. Casale, P., Pujol, O., Radeva, P.: Personalization and user verification in wearable systems using biometric walking patterns. Person. Ubiquitous Comput. **16**(5), 563–580 (2012)
25. Ordóñez, F.J., de Toledo, P., Sanchis, A.: Activity recognition using hybrid generative/discriminative models on home environments using binary sensors. Sensors **13**(5), 5460–5477 (2013)
26. Baños, O., Damas, M., Pomares, H., Rojas, I., Tóth, M.A., Amft, O.: A benchmark dataset to evaluate sensor displacement in activity recognition. In: Proceedings of the 2012 ACM Conference on Ubiquitous Computing, pp. 1026–1035. ACM (2012)
27. Reiss, A., Stricker, D.: Introducing a new benchmarked dataset for activity monitoring. In: 2012 16th International Symposium on Wearable Computers (ISWC), pp. 108–109. IEEE (2012)
28. Altun, K., Barshan, B., Tunçel, O.: Comparative study on classifying human activities with miniature inertial and magnetic sensors. Pattern Recognit. **43**(10), 3605–3620 (2010)
29. Bacciu, D., Barsocchi, P., Chessa, S., Gallicchio, C., Micheli, A.: An experimental characterization of reservoir computing in ambient assisted living applications. Neural Comput. Appl. **24**(6), 1451–1464 (2014)
30. Banos, O., Garcia, R., Holgado-Terriza, J.A., Damas, M., Pomares, H., Rojas, I., Saez, A., Villalonga, C.: Mhealthdroid: a novel framework for agile development of mobile health applications. In: International Workshop on Ambient Assisted Living, pp. 91–98. Springer (2014)
31. Weiss, G.M., Yoneda, K., Hayajneh, T.: Smartphone and smartwatch-based biometrics using activities of daily living. IEEE Access **7**, 133190–133202 (2019)
32. Schmidt, P., Reiss, A., Duerichen, R., Marberger, C., Van Laerhoven, K.: Introducing WESAD, a multimodal dataset for wearable stress and affect detection. In: Proceedings of the 2018 on International Conference on Multimodal Interaction, pp. 400–408. ACM (2018)
33. Özdemir, A., Barshan, B.: Detecting falls with wearable sensors using machine learning techniques. Sensors **14**(6), 10691–10708 (2014)
34. Bruno, B., Mastrogiovanni, F., Sgorbissa, A., Vernazza, T., Zaccaria, R.: Analysis of human behavior recognition algorithms based on acceleration data. In: 2013 IEEE International Conference on Robotics and Automation (ICRA), pp. 1602–1607. IEEE (2013)
35. Shoaib, M., Scholten, H., Havinga, P.J.M., Incel, O.D.: A hierarchical lazy smoking detection algorithm using smartwatch sensors. In: 2016 IEEE 18th International Conference on e-Health Networking, Applications and Services (Healthcom), pp. 1–6. IEEE (2016)
36. Shoaib, M., Bosch, S., Incel, O.D., Scholten, H., Havinga, P.J.M.: Complex human activity recognition using smartphone and wrist-worn motion sensors. Sensors **16**(4), 426 (2016)
37. Shoaib, M., Scholten, H., Havinga, P.J.M.: Towards physical activity recognition using smartphone sensors. In: Ubiquitous Intelligence and Computing, 2013 IEEE 10th International

Conference on and 10th International Conference on Autonomic and Trusted Computing (UIC/ATC), pp. 80–87. IEEE (2013)

38. Shoaib, M., Bosch, S., Incel, O.D., Scholten, H., Havinga, P.J.M.: Fusion of smartphone motion sensors for physical activity recognition. Sensors **14**(6), 10146–10176 (2014)

39. Hasc2010 corpus. http://hasc.jp. Accessed 27 Mar 2019

40. Kawaguchi, N., Yang, Y., Yang, T., Ogawa, N., Iwasaki, Y., Kaji, K., Terada, T., Murao, K., Inoue, S., Kawahara, Y. et al.: Hasc2011corpus: towards the common ground of human activity recognition. In: Proceedings of the 13th International Conference on Ubiquitous Computing, pp. 571–572. ACM (2011)

41. Kawaguchi, N., Watanabe, H., Yang, T., Ogawa, N., Iwasaki, Y., Kaji, K., Terada, T., Murao, K., Hada, H., Inoue, S., et al. Hasc2012corpus: large scale human activity corpus and its application. In: Proceedings of the Second International Workshop of Mobile Sensing: From Smartphones and Wearables to Big Data, pp. 10–14 (2012)

42. Kaji, K., Watanabe, H., Ban, R., Kawaguchi, N.: Hasc-ipsc: indoor pedestrian sensing corpus with a balance of gender and age for indoor positioning and floor-plan generation researches. In: Proceedings of the 2013 ACM Conference on Pervasive and Ubiquitous Computing Adjunct Publication, pp. 605–610. ACM (2013)

43. Ichino, H., Kaji, K., Sakurada, K., Hiroi, K., Kawaguchi, N.: Hasc-pac2016: large scale human pedestrian activity corpus and its baseline recognition. In: Proceedings of the 2016 ACM International Joint Conference on Pervasive and Ubiquitous Computing: Adjunct, pp. 705–714. ACM (2016)

44. Matsuyama, H., Hiroi, K., Kaji, K., Yonezawa, T., Kawaguchi, N.: Ballroom dance step type recognition by random forest using video and wearable sensor. In: Proceedings of the 2019 ACM International Joint Conference on Pervasive and Ubiquitous Computing and Proceedings of the 2019 ACM International Symposium on Wearable Computers, pp. 774–780. ACM (2019)

45. Hossain, T., Mairittha, T., Mairittha, N., Inoue, S., Lago, P.: Integrating activity recognition and nursing care records: the system, deployment, and a verification study. Proc. ACM Interact., Mob., Wear. Ubiquitous Technol. **3**(86) (2019)

46. Bachlin, M., Plotnik, M., Roggen, D., Maidan, I., Hausdorff, J.M., Giladi, N., Troster, G.: Wearable assistant for parkinson's disease patients with the freezing of gait symptom. IEEE Trans. Inf. Technol. Biomed. **14**(2), 436–446 (2010)

47. Inoue, S., Ueda, N., Nohara, Y., Nakashima, N.: Recognizing and understanding nursing activities for a whole day with a big dataset. J. Inf. Process. **24**(6), 853–866 (2016)

48. Predicting Parkinson's disease progression with smartphone data. https://www.kaggle.com/c/3300/download/Participant. Accessed 27 Mar 2019

49. Forster, K., Roggen, D., Troster, G.: Unsupervised classifier self-calibration through repeated context occurences: Is there robustness against sensor displacement to gain? In: International Symposium on Wearable Computers, 2009. ISWC'09, pp. 77–84. IEEE (2009)

50. Bächlin, M., Förster, K., Tröster, G.: Swimmaster: a wearable assistant for swimmer. In: Proceedings of the 11th International Conference on Ubiquitous Computing, pp. 215–224. ACM (2009)

51. Crowd-sourced fitbit datasets. Crowd-Sourced-Fitbit-Datasets (2016). Accessed 27 Mar 2019

52. Takata, M., Nakamura, Y., Fujimoto, M., Arakawa, Y., Yasumoto, K.: Investigating the effect of sensor position for training type recognition in a body weight training support system. In: Proceedings of the 2018 ACM International Joint Conference on Pervasive and Ubiquitous Computing and Proceedings of the 2018 ACM International Symposium on Wearable Computers, pp. 1–5. ACM (2018)

53. Alemdar, H., Ertan, H., Incel, O.D., Ersoy, C.: Aras human activity datasets in multiple homes with multiple residents. In: Proceedings of the 7th International Conference on Pervasive Computing Technologies for Healthcare, pp. 232–235. ICST (Institute for Computer Sciences, Social-Informatics and Telecommunications Engineering) (2013)

54. Tapia, E.M., Intille, S.S., Lopez, L., Larson, K.: The design of a portable kit of wireless sensors for naturalistic data collection. In: International Conference on Pervasive Computing, pp. 117–134. Springer, 2006

55. De la Torre, F., Hodgins, J., Bargteil, A., Martin, X., Macey, J., Collado, A., Beltran, P.: Guide to the Carnegie Mellon University Multimodal Activity (CMU-MMAC) Database. Robotics Institute, p. 135 (2008)
56. Chen, L., Nugent, C.D., Biswas, J., Hoey, J.: Activity Recognition in Pervasive Intelligent Environments, vol. 4. Springer Science & Business Media, Berlin (2011)
57. Gani, M.O., Saha, A.K., Ahsan, G.M.T., Ahamed, S.I., Smith, R.O.: A novel framework to recognize complex human activity. In: 2017 IEEE 41st Annual Computer Software and Applications Conference (COMPSAC), pp. 948–956. IEEE (2017)
58. Cook, D.J.: Learning setting-generalized activity models for smart spaces. IEEE Intell. Syst. **27**(1), 32–38 (2012)
59. Tapia, E.M., Intille, S.S., Larson, K.: Activity recognition in the home using simple and ubiquitous sensors. In: International Conference on Pervasive Computing, pp. 158–175. Springer (2004)
60. Huynh, T., Fritz, M., Schiele, B.: Discovery of activity patterns using topic models. In: Proceedings of the 10th International Conference on Ubiquitous Computing, pp. 10–19. ACM (2008)
61. Activity classification. https://www.kaggle.com. Accessed 28 Mar 2019
62. Eagle, N., Pentland, A.S.: Reality mining: sensing complex social systems. Person. Ubiquitous Comput. **10**(4), 255–268 (2006)
63. Laurila, J.K., Gatica-Perez, D., Aad, I., Bornet, O., Do, T.-M.-T., Dousse, O., Eberle, J., Miettinen, M. et al.: The mobile data challenge: Big data for mobile computing research. In: Pervasive Computing, Number EPFL-CONF-192489 (2012)
64. Wagner, D.T., Rice, A., Beresford, A.R.: Device analyzer: large-scale mobile data collection. ACM SIGMETRICS Perform. Eval. Rev. **41**(4), 53–56 (2014)
65. Rawassizadeh, R., Tomitsch, M., Nourizadeh, M., Momeni, E., Peery, A., Ulanova, L., Pazzani, M.: Energy-efficient integration of continuous context sensing and prediction into smartwatches. Sensors **15**(9), 22616–22645 (2015)
66. Zhang, M., Sawchuk, A.A.: USC-HAD: a daily activity dataset for ubiquitous activity recognition using wearable sensors. In: Proceedings of the 2012 ACM Conference on Ubiquitous Computing, pp. 1036–1043. ACM (2012)
67. Yang, A.Y., Jafari, R., Sastry, S.S., Bajcsy, R.: Distributed recognition of human actions using wearable motion sensor networks. J. Ambient Intell. Smart Environ. **1**(2), 103–115 (2009)
68. Stiefmeier, T., Roggen, D., Troster, G.: Fusion of string-matched templates for continuous activity recognition. In: 2007 11th IEEE International Symposium on Wearable Computers, pp. 41–44. IEEE (2007)
69. Wirz, M., Roggen, D., Troster, G.: Decentralized detection of group formations from wearable acceleration sensors. In: International Conference on Computational Science and Engineering, 2009. CSE'09, vol. 4, pp. 952–959. IEEE (2009)
70. Saha, S.S., Rahman, S., Rasna, M.J., Zahid, T.B., Mahfuzul Islam, A.K.M., Ahad, M.A.R.: Feature extraction, performance analysis and system design using the du mobility dataset. IEEE Access **6**, 44776–44786 (2018)
71. Saha, S.S., Rahman, S., Rasna, M.J., Mahfuzul Islam, A.K.M., Ahad, M.A.R., DU-MD: an open-source human action dataset for ubiquitous wearable sensors. In: Joint 7th International Conference on Informatics, Electronics and Vision; 2nd International Conference on Imaging, Vision and Pattern Recognition (2018)
72. Chereshnev, R., Kertész-Farkas, A.: Hugadb: Human gait database for activity recognition from wearable inertial sensor networks. In: International Conference on Analysis of Images, Social Networks and Texts, pp. 131–141. Springer (2017)
73. Chen, C., Jafari, R., Kehtarnavaz, N.: Utd-mhad: A multimodal dataset for human action recognition utilizing a depth camera and a wearable inertial sensor. In: 2015 IEEE International Conference on Image Processing (ICIP), pp. 168–172. IEEE (2015)
74. Ngo, T.T., Ahad, M.A.R., Antar, A.D., Ahmed, M., Muramatsu, D., Makihara, Y., Yagi, Y., Inoue, S., Hossain, T., Hattori, Y.: Ou-isir wearable sensor-based gait challenge: age and gender. In: Proceedings of the 12th IAPR International Conference on Biometrics, ICB (2019)

75. Antar, A.D., Ahmed, M., Hossain, T., Muramatsu, D., Makihara, Y., Inoue, S., Yagi, Y., Ahad, M.A.R., Ngo, T.T.: Wearable sensor-based gait analysis for age and gender estimation (2020)
76. Kwapisz, J.R., Weiss, G.M., Moore, S.A.: Activity recognition using cell phone accelerometers. ACM SigKDD Explor. Newsl. **12**(2), 74–82 (2011)
77. Faye, S., Louveton, N., Jafarnejad, S., Kryvchenko, R., Engel, T.: An open dataset for human activity analysis using smart devices (2017)
78. Ngo, T.T., Makihara, Y., Nagahara, H., Mukaigawa, Y., Yagi, Y.: The largest inertial sensor-based gait database and performance evaluation of gait-based personal authentication. Pattern Recognit. **47**(1), 228–237 (2014)
79. Gjoreski, H., Kaluža, B., Gams, M., Milić, R., Luštrek, M.: Context-based ensemble method for human energy expenditure estimation. Appl. Soft Comput. **37**, 960–970 (2015)
80. Suhr, J.K., Jung, H.G.: Automatic parking space detection and tracking for underground and indoor environments. IEEE Trans. Indust. Electron. **63**(9), 5687–5698 (2016)
81. Mekki, S., Karagkioules, T., Valentin, S.: Context-aware adaptive video streaming for mobile users. In: 2017 IEEE Conference on Computer Communications Workshops (INFOCOM WKSHPS), pp. 988–989. IEEE (2017)
82. Mekki, S., Karagkioules, T.: Http adaptive streaming with indoors-outdoors detection in mobile networks. In: 2017 IEEE Conference on Computer Communications Workshops (INFOCOM WKSHPS), pp. 671–676. IEEE (2017)
83. Gjoreski, H., Ciliberto, M., Wang, L., Morales, F.J.O., Mekki, S., Valentin, S., Roggen, D.: The university of sussex-huawei locomotion and transportation dataset for multimodal analytics with mobile devices. IEEE Access (2018)
84. Ahmed, M., Antar, A.D., Hossain, T., Inoue, S., Ahad, M.A.R.: Poiden: position and orientation independent deep ensemble network for the classification of locomotion and transportation modes. pp. 674–679 (2019)
85. Antar, A.D., Ahmed, M., Ishrak, M.S., Ahad, M.A.R.: A comparative approach to classification of locomotion and transportation modes using smartphone sensor data. In: Proceedings of the 2018 ACM International Joint Conference and 2018 International Symposium on Pervasive and Ubiquitous Computing and Wearable Computers, pp. 1497–1502 (2018)
86. Rasna, M.J., Hossain, T., Inoue, S., Sha, S.S., Rahman, S., Ahad, M.A.R.: Supervised and neural classifiers for locomotion analysis. 2018 ACM International Joint Conference on Pervasive and Ubiquitous Computing and the 2018 International Symposium on Wearable Computers (UbiComp/ISWC) (2018)
87. Yang, J.: Toward physical activity diary: motion recognition using simple acceleration features with mobile phones. In: Proceedings of the 1st International Workshop on Interactive Multimedia for Consumer Electronics, pp. 1–10. ACM (2009)
88. Zheng, Y., Xie, X., Ma, W.-Y.: Geolife: a collaborative social networking service among user, location and trajectory. IEEE Data Eng. Bull. **33**(2), 32–39 (2010)
89. Wang, S., Chen, C., Ma, J.: Accelerometer based transportation mode recognition on mobile phones. In: 2010 Asia-Pacific Conference on Wearable Computing Systems (APWCS), pp. 44–46. IEEE (2010)
90. Reddy, S., Mun, M., Burke, J., Estrin, D., Hansen, M., Srivastava, M.: Using mobile phones to determine transportation modes. ACM Trans. Sens. Netw. (TOSN) **6**(2), 13 (2010)
91. Siirtola, P., Röning, J.: Recognizing human activities user-independently on smartphones based on accelerometer data. IJIMAI **1**(5), 38–45 (2012)
92. Hemminki, S., Nurmi, P., Tarkoma, S.: Accelerometer-based transportation mode detection on smartphones. In: Proceedings of the 11th ACM Conference on Embedded Networked Sensor Systems, p. 13. ACM (2013)
93. Zhang, Z., Poslad, S.: A new post correction algorithm (pocoa) for improved transportation mode recognition. In: 2013 IEEE International Conference on Systems, Man, and Cybernetics (SMC), pp. 1512–1518. IEEE (2013)
94. Xia, H., Qiao, Y., Jian, J., Chang, Y.: Using smart phone sensors to detect transportation modes. Sensors **14**(11), 20843–20865 (2014)

95. Yu, M.-C., Yu, T., Wang, S.-C., Lin, C.-J., Chang, E.Y.: Big data small footprint: the design of a low-power classifier for detecting transportation modes. Proc. VLDB Endow. **7**(13), 1429–1440 (2014)
96. Widhalm, P., Nitsche, P., Brändie, N.: Transport mode detection with realistic smartphone sensor data. In: 2012 21st International Conference on Pattern Recognition (ICPR), pp. 573–576. IEEE (2012)
97. Jahangiri, A., Rakha, H.A.: Applying machine learning techniques to transportation mode recognition using mobile phone sensor data. IEEE Trans. Intell. Transp. Syst. **16**(5), 2406–2417 (2015)
98. Xing, S., Caceres, H., Tong, H., He, Q.: Online travel mode identification using smartphones with battery saving considerations. IEEE Trans. Intell. Transp. Syst. **17**(10), 2921–2934 (2016)
99. Gjoreski, H., Ciliberto, M., Wang, L., Morales, F.J.O., Mekki, S., Valentin, S., Roggen, D.: The university of sussex-huawei locomotion and transportation dataset for multimodal analytics with mobile devices. IEEE Access **6**, 42592–42604 (2018)
100. Islam, M.Z., Serikawa, S., Islam, Z.Z., Tazwar, S.M., Ahad, M.A.R.: Automatic fall detection system of unsupervised elderly people using smartphone. In: Annual Conference on Artificial Intelligence. IEEE (2017)
101. Casilari, E., Santoyo-Ramón, J.-A., Cano-García, J.-M.: Analysis of public datasets for wearable fall detection systems. Sensors **17**(7), 1513 (2017)
102. Frank, K., Nadales, M.J.V., Robertson, P., Pfeifer, T.: Bayesian recognition of motion related activities with inertial sensors. In: Proceedings of the 12th ACM International Conference Adjunct Papers on Ubiquitous Computing-Adjunct, pp. 445–446. ACM (2010)
103. Vavoulas, G., Pediaditis, M., Spanakis, E.G., Tsiknakis, M.: The mobifall dataset: an initial evaluation of fall detection algorithms using smartphones. In: 2013 IEEE 13th International Conference on Bioinformatics and Bioengineering (BIBE), pp. 1–4. IEEE (2013)
104. Kwolek, B., Kepski, M.: Human fall detection on embedded platform using depth maps and wireless accelerometer. Comput. Methods Programs Biomed. **117**(3), 489–501 (2014)
105. Medrano, C., Igual, R., Plaza, I., Castro, M.: Detecting falls as novelties in acceleration patterns acquired with smartphones. PloS one **9**(4), e94811 (2014)
106. Gasparrini, S., Cippitelli, E., Spinsante, S., Gambi, E.: A depth-based fall detection system using a kinect® sensor. Sensors **14**(2), 2756–2775 (2014)
107. Ojetola, O., Gaura, E., Brusey, J.: Data set for fall events and daily activities from inertial sensors. In: Proceedings of the 6th ACM Multimedia Systems Conference, pp. 243–248. ACM (2015)
108. Vilarinho, T., Farshchian, B., Bajer, D.G., Dahl, O.H., Egge, I., Hegdal, S.S., Lønes, A., Slettevold, J.N., Weggersen, S.M.: A combined smartphone and smartwatch fall detection system. In: 2015 IEEE International Conference on Computer and Information Technology; Ubiquitous Computing and Communications; Dependable, Autonomic and Secure Computing; Pervasive Intelligence and Computing (CIT/IUCC/DASC/PICOM), pp. 1443–1448. IEEE (2015)
109. Micucci, D., Mobilio, M., Napoletano, P.: Unimib shar: a dataset for human activity recognition using acceleration data from smartphones. Appl. Sci. **7**(10), 1101 (2017)
110. Casilari, E., Santoyo-Ramón, J.A., Cano-García, J.M.: Analysis of a smartphone-based architecture with multiple mobility sensors for fall detection. PLoS one **11**(12), e0168069 (2016)
111. Vavoulas, G., Chatzaki, C., Malliotakis, T., Pediaditis, M., Tsiknakis, M.: The mobiact dataset: recognition of activities of daily living using smartphones. In: ICT4AgeingWell, pp. 143–151 (2016)
112. Sucerquia, A., López, J.D., Vargas-Bonilla, J.F.: Sisfall: a fall and movement dataset. Sensors **17**(1), 198 (2017)
113. Chen, Y., Keogh, E., Hu, B., Begum, N., Bagnall, A., Mueen, A., Batista, G.: The ucr time series classification archive (2015). www.cs.ucr.edu/eamonn/time_series_data
114. Goldberger, A.L.: Physiobank, physiotoolkkit, and physionet: components of a new research resource for complex physiologic signals. Circulation **101**(23), e215–e220 (2000)

115. Kotz, D., Henderson, T.: Crawdad: a community resource for archiving wireless data at dartmouth. IEEE Pervas. Comput. **4**(4), 12–14 (2005)
116. Syeda, U.H., Zafar, Z., Islam, Z.Z., Tazwar, S.M., Rasna, M.J., Kise, K., Ahad, M.A.R.: Visual face scanning and emotion perception analysis between autistic and typically developing children. In: ACM UbiComp Workshop on Mental Health and Well-being: Sensing and Intervention. ACM (2017)
117. Rahaman, N., Islam, A., Ahad, M.A.R.: A study on tiredness assessment by using eye blink detection, pp. 209–214 (2019)
118. Noman, M.T.N., Ahad, M.A.R.: Mobile-based eye-blink detection performance analysis on android platform (2018)
119. Kawsar, F., Min, C., Mathur, A., Montanari, A.: Earables for personal-scale behavior analytics. IEEE Pervas. Comput. **17**(3), 83–89 (2018)
120. Noman, M.T.N., Hussein, M.A.H., Ahad, M.A.R.: A study on tiredness measurement using computer vision, pp. 110–117 (2019)
121. Irtija, M.S.N., Ahad, M.A.R.: Fatigue detection using facial landmarks. In: 4th International Symposium on Affective Science and Engineering, and the 29th Modern Artificial Intelligence and Cognitive Science Conference (ISASE-MAICS) (2018)

# Chapter 7
# An Overview of Classification Issues in Sensor-Based Activity Recognition

**Abstract** Human activity analysis and recognition tasks can be considered as classification problems in most of the cases. This chapter represents the overview of classification problems explaining their different types with examples. Binary classification, multilabel classification, multi-class classification, and hierarchical classification tasks are presented in this chapter. At the end, we have given a brief summary about the bias-variance trade-off problem.

## 7.1 Types of Classification Problems

Sensor-based Human Activity Recognition (HAR) has been explored by many research communities and industries for various applications—along with various challenges ahead to deal with [1–8]. Sensor-based activity recognition has numerous application domains that widen the classification tasks into a great extent. In previous research works, human activity recognition was restricted into binary classification and multi-class classification tasks. Due to the increasing application tasks in the medical domain, multilabel and hierarchical classification tasks have also been significant nowadays. In general, there can be four types of classification problems as shown in Fig. 7.1:

- Binary classification,
- Multi-class classification,
- Multilabel classification, and
- Hierarchical classification.

Different research works exploited different classifiers [9–13]. We discuss these classification tasks based on their definitions and use cases.

© Springer Nature Switzerland AG 2021  
M. A. R. Ahad et al., *IoT Sensor-Based Activity Recognition*, Intelligent Systems Reference Library 173, https://doi.org/10.1007/978-3-030-51379-5_7

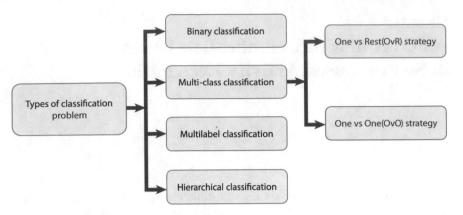

**Fig. 7.1**    Different types of classification problems

## 7.2    Binary Classification

Binary classification is also known as binomial classification where the task is to classify the components of an assigned set into two groups following a classification rule. In a binary classification task, the input is generally classified into one of the two non-overlapping classes. Non-overlapping classes signify that the classes must be mutually exclusive. This intends that there is no way that any of the data could fall into two different classes at once. For example, utilizing heartbeat sensor and body pressure data to evaluate whether a patient has a certain disease or not. Some of the well-known methods that are used for the binary classification task are,

- Decision trees,
- Random forests,
- Neural networks,
- Logistic regression,
- Bayesian networks,
- Support Vector Machines (SVM),
- Probit model, etc.

Binary classification technique has been used in many research domains. Binary-classification experiments have been discussed in [14], where response time was the primary performance measure to review some studies of human information-processing capabilities. The feature subset selection problem has been analyzed in [15] for the binary classification problem using a logistic regression model. There is another research work where the authors have investigated the polarity correspondence principle for the performance of speeded binary classification tasks [16].

# 7.3   Multi-class Classification

Multi-class classification is also known as multinomial classification where the task is to classify the instances of input data into one of three or more classes. In this case, the input is classified into one of the $n$ non-overlapping classes. The value of $n$ should be equal to or greater than 3 for multi-class classification. For example, classifying human activities (e.g., walking, jogging, running, upstairs, downstairs, sitting, etc.) using a smartphone or smartwatch sensor (e.g., accelerometer and/or gyroscope) data can be called multi-class classification task. Some of the algorithms to approach multi-class classification problems are,

- Neural networks,
- Extreme learning machines (a special case of single hidden layer feed-forward neural networks),
- K-nearest neighbors (KNN),
- Naive Bayes,
- Decision trees,
- Support Vector Machines (SVM), etc.

As human activities are complex and there can be many variations of activities, multi-class classification is the common choice for human activity recognition tasks. Multiclass Support Vector Machine algorithm has been used for human physical Activity Recognition using acceleration sensors [17–23]. Boosting algorithms can also be used for multi-class classification of human activities [24]. Tree-based ensemble classifiers for multiclass-classification are also common to recognize and classify human activities with great performance [9–12, 25].

These types of multi-class classification methods are also known as algorithm adaptation techniques. There is another strategy to decrease the problem of multi-class classification by transforming it into multiple binary classification problems. This method can be called a problem transformation technique that can be utilized in two ways:

- One versus Rest (OvR) strategy
- One versus One (OvO) strategy

## 7.3.1   One Versus Rest (OvR) Strategy

This method is also named as One versus All (OvA) and One Against All (OAA) strategy. This method is also known as binary relevance method. In this case, a single classifier is trained for each class and it is considered that only the samples of that class belong to the positive class and all other samples belong to the negative class. This is required to produce a real-valued confidence score by the base classifier rather than just a class label for its decision. The class is fitted against all other classes for each classifier in this method.

This method is computationally efficient and interpretable for a fewer amount of activity classes as the total number of required classifiers is equal to the number of classes. Since only one classifier represents one class, it is possible to gain knowledge about that class by inspecting only a single classifier output. In spite of these advantages, this strategy has several drawbacks:

- First of all, the binary classifiers may differ in terms of the scale of confidence values.
- Secondly, even if it is possible to create a training set with balanced class distributions, the binary classification learners see unbalanced distribution because the set of negatives they see is much higher than the set of positives.

### 7.3.2  One Versus One (OvO) Strategy

In this method, one classifier is fitted per class pair. Each classifier takes the samples of a pair of classes from the initial training set and must learn to recognize these two classes. A voting scheme is applied for all of the classifiers at the prediction time. During the prediction time, the class with the most received vote is selected as the output class. So, it requires $\frac{n(n-1)}{2}$ classifiers in total, where, $n$ = total number of classes. This method has a higher amount of computational cost than the One versus Rest method because of the higher amount of classifiers for prediction. However, this strategy has benefits for kernel algorithms, which do not scale well with a higher number of samples. The reason behind this is the individual learning problem that only requires a small subset of the data whereas, with One versus Rest strategy, the complete dataset is used for $n$ times, where $n$ is the number of classes.

## 7.4  Multilabel Classification

Though multiclass and multilabel classification tasks sound similar, there is a clear difference in terms of the number of predicted labels for each instance. In the case of multilabel classification, multiple labels are to be predicted for each instance [26]. In a formal manner, in the task of multilabel classification, a model is required to design that maps inputs $x$ to binary vectors $y$, assigning 0 or 1 to each element of output vectors. The classes are mutually exclusive for the multiclass classification task, whereas, for multilabel classification, each label depicts a different classification task, but the tasks are somehow related.

For example, we can take data from an eye tracker while watching a movie and our task is to classify the genre (e.g., horror, romance, adventure, documentary, comedy, and science fiction) of the movie based on the eye tracker data. These classes are not mutually exclusive, as a movie can be classified into more than one class (for example, horror and adventure, romantic and horror, documentary, horror, and adventure, etc.)

**Table 7.1** Multilabel classification task of finding genres of a movie by utilizing eye tracker data

| Movie genre | Percentage of belonging To a particular class (in %) |
|---|---|
| Comedy | 0.7 |
| Horror | 0.02 |
| Romance | 90.08 |
| Adventure | 6.1 |
| Documentary | 0 |
| Science fiction | 2.2 |

with varying percentages. Therefore, this is a multilabel classification problem. The results from Table 7.1 summarizes this situation (This table has been generated using random values). In the case of multiclass classification, on the other hand, each instance can be assigned to only one label (for example, the activity can be either *walk* or *sit* but cannot be both at the same time).

Some algorithms that can be adapted for the multilabel classification task are,

- Boosting,
- K-nearest neighbors (KNN),
- Decision trees,
- Kernel methods for vector output,
- Neural networks.

Multilabel classification technique has been used to train human activity recognition algorithms for labels with inaccurate time stamps [27]. It can also be used for physical activity recognition from accelerometer sensor values from different positions [28]. There are some other research works in this domain where multilabel classification technique has been used [29–32].

## 7.5 Hierarchical Classification

In this method, the input is classified into only one class, which can be divided into subclasses or grouped into superclasses. The hierarchy needs to be predefined and cannot be changed while performing the classification task. In this case, firstly, the classification is performed at a lower level with highly specific input data. Then it is required to systematically combine the classifications of the individual pieces of data and classified on a higher level iteratively until one output is produced. This final output is the classification of the data overall. This method can be useful for recognizing pictures in the area of computer vision. There are other applications in the field of text classification [33] and protein function prediction.

The hierarchical classification technique is very popular in computer vision, where human activities can be identified from a video [34, 35]. However, this technique can also be applied to sensor-based human activity recognition [36]. Hierarchical Hidden Markov Model is popular in this case [37]. There are also some other research works in this domain focused on hierarchical modeling for human activity recognition [38–41].

## 7.6  Bias-Variance Trade-Off

In this section, we have discussed the bias-variance trade-off problem, which can be used to properly interpret different machine learning algorithms and to evade the mistake of overfitting and underfitting.

### 7.6.1  Bias Error

Bias signifies the difference between the average prediction of our model and the correct value, which we are trying to predict. Highly biased models do not care much about training data and make simplified assumptions. In general, linear algorithms are highly biased, which makes the learning process faster but they are less flexible. In the case of complex problems, biased models have lower predictive performance in general [42]. So we can say that low bias models, for example, decision trees, k-nearest neighbors, Support Vector Machine, etc. suggest less assumption about the form of the target function, whereas, highly biased models i.e., Linear Regression, Linear Discriminant Analysis, Logistic Regression, etc. suggest more assumption about the form of the target function.

### 7.6.2  Variance Error

Variance defines how much model prediction varies for different training data, which also tells us the spread of our data. Models with high variance try to learn the training data well and do not generalize well on the test data. They are heavily influenced by the specifics of the training data. In general, non-linear machine learning algorithms with high flexibility (i.e., decision tree) have high variance. Machine learning algorithms with low variance, for example, Linear Regression, Linear Discriminant Analysis, Logistic Regression, etc. suggest minute adjustments to the estimate of the target function with changes to the training dataset. On the other hand, machine learning algorithms with high variance i.e., decision rees, k-nearest neighbors, Support Vector Machines, etc. suggest large changes to the estimate of the target function with changes to the training dataset.

### 7.6.3   Irreducible Error

There is another error named irreducible error, which can not be overcome notwithstanding of what algorithm is used. This is the error that stemmed from the problem's preferred representation. Factors such as unknown variables that influence input variables mapping to output variable can trigger this error [42].

### 7.6.4   Overfitting and Underfitting Related to Bias and Variance

From the discussion above, it is clear that the goal of supervised learning algorithms is to obtain low bias and low variance to achieve good prediction performance. From another point of view, in the case of models that underfit the data and fail to capture the underlying pattern, they usually have high bias and low variance. This case is prominent while using a lower amount of data to build a model where the model is too simple to capture the complex pattern in the data or in the case of building linear models with nonlinear data. On the other hand, when the models overfit capturing noise along with the underlying pattern of the data, these models usually have low bias and high variance. This situation appears when we train our model over a noisy dataset. In general, the complex models are prone to overfitting with high variance.

### 7.6.5   Balancing Bias-Variance Trade-Off

Too simple models with very few parameters have high bias and low variance, whereas, complex models with a large number of parameters have high variance and low bias. An algorithm can not be more complex and less complex at the same time. Increasing the bias will decrease the variance and increasing the variance will decrease the bias. So, there is a trade-off between these two concerns and we need to find a balance between in this trade-off by tuning the model hyperparameters. To build a reliable model, we necessitate finding a suitable balance between bias and variance such that it minimizes the total error. The $Total Error$ can be defined as,

$$Total\ Error = Bias^2 + Variance + Irreducible\ Error \qquad (7.1)$$

Therefore, to understand the behavior of prediction models, it is really important to understand the concept of bias and variance. If we can create an optimal balance of bias and variance, it would never overfit or underfit the model.

## 7.7  Conclusion

In order to obtain the desired performance in classification problems, we need to categorize our problem into a suitable classification task. Then we need to know about that classification category in detail. Considering these tasks, we have designed this chapter that provides a summary of different types of classification tasks with proper explanation and examples. We have also compared the One versus One and One versus Rest classification tasks in this chapter. We have also discussed the problem of bias-variance trade-off to find a good balance between bias and variance such that it minimizes the total error. This will help to build a good model.

## 7.8  Think Further

1. What are the different types of classification problems?
2. What is binary classification?
3. In which cases we can use binary classification?
4. What is multiclass classification?
5. In which cases we can use multiclass classification?
6. Differentiate between one versus rest (OvR) and one versus one (OvO) strategy.
7. What is multilabel classification?
8. In which cases we can use multilabel classification?
9. What is hierarchical classification?
10. In which cases we can use hierarchical classification?
11. What is Bias and Variance?
12. Why can not we eliminate irreducible error easily?
13. How to deal with bias-variance trade-off?

## References

1. Antar, A.D., Ahad, M.A.R., Shahid, O.: Vision-based action understanding for assistive health-care: a short review. IEEE CVPR Workshop (2019)
2. Ahad, M.A.R.: Vision and sensor based human activity recognition: Challenges Ahead (2020)
3. Antar, A.D., Ahmed, M., Ahad, M.A.R.: Challenges in sensor-based human activity recognition and a comparative analysis of benchmark datasets: a review. In: 2019 Joint 8th International Conference on Informatics, Electronics and Vision (ICIEV) and 2019 3rd International Conference on Imaging, Vision and Pattern Recognition (icIVPR), pp. 134–139. IEEE (2019)
4. Ahad, M.A.R.: Motion History Images for Action Recognition and Understanding. Springer Science & Business Media, Berlin (2012)
5. Ahad, M.A.R.: Computer Vision and Action Recognition: a Guide for Image Processing and Computer Vision Community for Action Understanding, vol. 5. Springer Science & Business Media, Berlin (2011)

6. Hossain, T., Islam M.S., Ahad, M.A.R., Inoue, S.: Human activity recognition using earable device. In: Proceedings of the 2019 ACM International Joint Conference on Pervasive and Ubiquitous Computing and Proceedings of the 2019 ACM International Symposium on Wearable Computers, pp. 81–84. ACM (2019)
7. Tazin, T., Hossain, T., Ahad, M.A.R., Inoue, S.: Activity recognition by using lorawan sensor. In: 2018 ACM International Joint Conference on Pervasive and Ubiquitous Computing and the 2018 International Symposium on Wearable Computers (UbiComp/ISWC) (2018)
8. Ahmed, M., Antar, A.D., Ahad, M.A.R.: An approach to classify human activities in real-time from smartphone sensor data. In: 2019 Joint 8th International Conference on Informatics, Electronics Vision (ICIEV) and 2019 3rd International Conference on Imaging, Vision Pattern Recognition (icIVPR), pp. 140–145 (2019)
9. Uddin, M.T., Uddiny, M.A.: Human activity recognition from wearable sensors using extremely randomized trees. In: 2015 International Conference on Electrical Engineering and Information Communication Technology (ICEEICT), pp. 1-6. IEEE (2015)
10. Ahmed, M., Antar, A.D., Hossain, T., Inoue, S., Ahad, M.A.R.: Poiden: position and orientation independent deep ensemble network for the classification of locomotion and transportation modes. pp. 674–679 (2019)
11. Antar, A.D., Ahmed, M., Ishrak M.S., Ahad, M.A.R.: A comparative approach to classification of locomotion and transportation modes using smartphone sensor data. In: Proceedings of the 2018 ACM International Joint Conference and 2018 International Symposium on Pervasive and Ubiquitous Computing and Wearable Computers, pp. 1497–1502 (2018)
12. Rasna, M.J., Hossain, T., Inoue, S., Sha, S.S., Rahman, S., Ahad, M.A.R.: Supervised and neural classifiers for locomotion analysis. 2018 ACM International Joint Conference on Pervasive and Ubiquitous Computing and the 2018 International Symposium on Wearable Computers (UbiComp/ISWC) (2018)
13. Ngo, T.T., Ahad, M.A.R., Antar, A.D., Ahmed, M., Muramatsu, D., Makihara, Y., Yagi, Y., Inoue, S., Hossain, T., Hattori, Y.: Ou-isir wearable sensor-based gait challenge: age and gender. In: Proceedings of the 12th IAPR International Conference on Biometrics, ICB (2019)
14. Nickerson, R.S.: Binary-classification reaction time: a review of some studies of human information-processing capabilities. In: Psychonomic Monograph Supplements (1972)
15. Unler, A., Murat, A.: A discrete particle swarm optimization method for feature selection in binary classification problems. Eur. J. Oper. Res. **206**(3), 528–539 (2010)
16. Proctor, R.W., Cho, Y.S.: Polarity correspondence: a general principle for performance of speeded binary classification tasks. Psychol. Bull. **132**(3), 416 (2006)
17. He, Z., Jin, L.: Activity recognition from acceleration data based on discrete consine transform and svm. In: 2009 IEEE International Conference on Systems, Man and Cybernetics, pp. 5041–5044. IEEE (2009)
18. Anguita, D., Ghio, A., Oneto, L., Parra, X., Reyes-Ortiz, J.L.: Human activity recognition on smartphones using a multiclass hardware-friendly support vector machine. In: International Workshop on Ambient Assisted Living, pp. 216–223. Springer (2012)
19. Anguita, D., Ghio, A., Oneto, L., Parra, X., Reyes-Ortiz, J.L.: A public domain dataset for human activity recognition using smartphones. In: Esann (2013)
20. Althloothi, S., Mahoor, M.H., Zhang, X., Voyles, R.M.: Human activity recognition using multi-features and multiple kernel learning. Pattern Recognit. **47**(5), 1800–1812 (2014)
21. Palaniappan, A., Bhargavi, R., Vaidehi, V.: Abnormal human activity recognition using svm based approach. In: 2012 International Conference on Recent Trends in Information Technology, pp. 97–102. IEEE (2012)
22. Manosha Chathuramali, K.G., Rodrigo, R.: Faster human activity recognition with svm. In: International Conference on Advances in ICT for Emerging Regions (ICTer2012), pp. 197–203. IEEE (2012)
23. Qian, H., Mao, Y., Xiang, W., Wang, Z.: Recognition of human activities using svm multi-class classifier. Pattern Recognit. Lett. **31**(2), 100–111 (2010)

24. Reiss, A., Hendeby, G., Stricker, D.: A competitive approach for human activity recognition on smartphones. In: European Symposium on Artificial Neural Networks, Computational Intelligence and Machine Learning (ESANN 2013), 24–26 April, Bruges, Belgium, pp. 455–460. ESANN (2013)

25. Ghose, S., Mitra, J., Karunanithi, M., Dowling, J.: Human activity recognition from smartphone sensor data using a multi-class ensemble learning in home monitoring. Stud. Health Technol. Inform **214**, 62–67 (2015)

26. Tikk, D., Biró, G.: Experiments with multi-label text classifier on the reuters collection. In: Proceedings of the international conference on computational cybernetics (ICCC 03), pp. 33–38 (2003)

27. Toda, T., Inoue, S., Tanaka, S., Ueda, N.: Training human activity recognition for labels with inaccurate time stamps. In: Proceedings of the 2014 ACM International Joint Conference on Pervasive and Ubiquitous Computing: Adjunct Publication, pp. 863–872 (2014)

28. Mohamed, R.: Multi-label classification for physical activity recognition from various accelerometer sensor positions. J. Inf. Commun. Technol. **17**(2), 209–231 (2020)

29. Mosabbeb, E.A., Cabral, R., De la Torre, F., Fathy, M.: Multi-label discriminative weakly-supervised human activity recognition and localization. In: Asian Conference on Computer Vision, pp. 241–258. Springer (2014)

30. Kumar, R., Qamar, I., Virdi, J.S., Krishnan, N.C.: Multi-label learning for activity recognition. In: 2015 International Conference on Intelligent Environments, pp. 152–155. IEEE (2015)

31. Mohamed, R., Perumal, T., Sulaiman, M., Mustapha, N., Zainudin, M.N., et al.: Multi label classification on multi resident in smart home using classifier chains. Adv. Sci. Lett **24**(2), 1316–1319 (2018)

32. Cheema, M.S., Eweiwi, A., Bauckhage, C.: Human activity recognition by separating style and content. Pattern Recognit. Lett. **50**, 130–138 (2014)

33. Sun, A., Lim, E.-P., Ng, W.-K.: Performance measurement framework for hierarchical text classification. J. Am. Soc. Inf. Sci. Technol. **54**(11), 1014–1028 (2003)

34. Ribeiro, P.C., Santos-Victor, J., Lisboa, P.: Human activity recognition from video: modeling, feature selection and classification architecture. In: Proceedings of International Workshop on Human Activity Recognition and Modelling, pp. 61–78. Citeseer (2005)

35. Yin, J., Meng, Y.: Human activity recognition in video using a hierarchical probabilistic latent model. In: 2010 IEEE Computer Society Conference on Computer Vision and Pattern Recognition-Workshops, pp. 15–20. IEEE (2010)

36. Banos, O., Damas, M., Pomares, H., Rojas, F., Delgado-Marquez, B., Valenzuela, O.: Human activity recognition based on a sensor weighting hierarchical classifier. Soft Comput. **17**(2), 333–343 (2013)

37. Incel, O.D., Kose, M., Ersoy, C.: A review and taxonomy of activity recognition on mobile phones. BioNanoScience **3**(2), 145–171 (2013)

38. Khan, A.M., Lee, Y.-K., Lee, S.Y., Kim, T.-S.: A triaxial accelerometer-based physical-activity recognition via augmented-signal features and a hierarchical recognizer. IEEE Trans. Inf. Technol. Biomed. **14**(5), 1166–1172 (2010)

39. Lan, T., Sigal, L., Mori, G.: Social roles in hierarchical models for human activity recognition. In: 2012 IEEE Conference on Computer Vision and Pattern Recognition, pp. 1354–1361. IEEE (2012)

40. Duong, T., Phung, D., Bui, H., Venkatesh, S.: Efficient duration and hierarchical modeling for human activity recognition. Artif. Intell. **173**(7–8), 830–856 (2009)

41. Yuhuang, Z.: Human activity recognition based on the hierarchical feature selection and classification framework. J. Electrical Comput. Eng. **2015**, (2015)

42. Brownlee, J.: Master machine learning algorithms: discover how they work and implement them from scratch. In: Machine Learning Mastery (2016)

# Chapter 8
# Performance Evaluation in Activity Classification: Factors to Consider

**Abstract** After building the model to recognize activities from sensor data, it is essential to investigate the effectiveness of the model. The evaluation of the performance for machine learning methods can be performed using some evaluation matrices. This chapter properly explains the evaluation matrices namely accuracy, precision, recall, F1 score, balance classification rate, confusion matrix, and so on. Graphical performance measures namely ROC curve, cumulative gains, and lift charts have been explained too. This chapter also represents some essential concepts related to precision and recall trade-off, and accuracy as a performance measure. The contents of this chapter will be useful not only for human activity recognition, but also for other classification-related researches.

## 8.1 Performance Measure

Sensor-based Human Activity Recognition (HAR) has been explored by many research communities and industries for various applications—along with various challenges ahead to deal with [1–8]. In this chapter, we present various performance evaluation issues in human activity recognition. Proper evaluation and performance analysis of machine learning methods and models that have been utilized or proposed in research works is an essential part that must be taken care of. Though a specific model can produce good results in terms of accuracy-based evaluation, it may happen that the same model is performing poor while evaluated using logarithmic loss or other matrices. This phenomenon shows the importance of choosing proper evaluation matrices for the performance analysis of the proposed models in sensor-based activity recognition. Most common matrices are –

- Accuracy,
- Recall,
- Precision,
- F1 score, and
- ROC curve.

© Springer Nature Switzerland AG 2021

M. A. R. Ahad et al., *IoT Sensor-Based Activity Recognition*, Intelligent Systems Reference Library 173, https://doi.org/10.1007/978-3-030-51379-5_8

These can be categorized under paired criteria, combined criteria, and graphical tools [9–15]. Research has also been done to make the classification technique robust by analyzing the performance properly in imprecise environments [16].

## 8.2   Four Essential Concepts of Classification

Four terms are closely related to measure the performance of a model while predicting a class. Every evaluation matrices are dependent on these four terms—based on True or False, and Positive or Negative. These four concepts are:

- True positives
- True negatives
- False positives
- False negatives

### 8.2.1   True Positive ($T_P$)

True positive denotes the number of outcomes where the model correctly predicts the positive class. This refers to the positive tuples correctly labeled by the classifier. For example, if an umpire gives a batsman not out when he is not out, this is a true positive case.

### 8.2.2   True Negative ($T_N$)

True negative indicates the number of outcomes where the model correctly predicts the negative class. These are the negative tuples that are correctly identified by the classifier. If an umpire gives a batsman out when he is out, this can be an example of a true negative case.

### 8.2.3   False Positive ($F_P$)

False Positive denotes the number of outcomes where the model *incorrectly* predicts the positive class. These are the negative tuples that are incorrectly labeled as positive. For example, if an umpire gives a batsman not out when he is out. False positive is also known as *type-1 error*.

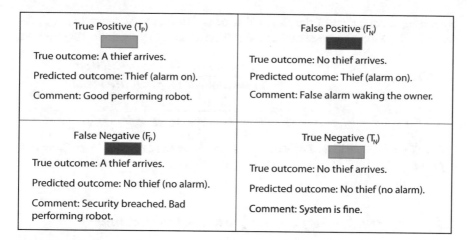

**Fig. 8.1** Concept of true positive, true negative, false positive, and false negative in the case of a thief predictor model using environmental sensors

## 8.2.4 False Negative ($F_N$)

False Negative represents the number of outcomes where the model *incorrectly* predicts the negative class. These are the positive tuples that are incorrectly labeled as negative. If an umpire gives a batsman out when he is not out. This error is the most severe during classification and it is regarded as a *type-2 error*.

## 8.2.5 Example to Clear the Concept

In this section, we discuss to clear the concept of true positive, true negative, false positive, and false negative by providing an example of a robot with motion sensors, light sensors, etc. for alarming when a thief attacks a home. Let's assume that a thief arrives, this is a positive class, and no thief arrives, this is a negative class. We can summarize the thief prediction model by using a 2 × 2 confusion matrix as shown in Fig. 8.1, which depicts four possible outcomes.

## 8.3 Classification Accuracy

Classification accuracy is defined as the ratio of the number of correct prediction by a model to the number of input samples in total. Based on our definition of true positive, true negative, false positive, and false negative, we can deduce that the total number of correct prediction is the sum of true positive and true negative, whereas the total input sample comprises of the summation of the all four parameters ($T_P$, $T_N$, $F_P$, and $F_N$). So, we can represent the classification accuracy by the following formula,

$$Classification\ Accuracy = \frac{Number\ of\ correct\ prediction}{Total\ number\ of\ prediction\ made}$$
$$= \frac{T_P + T_N}{T_P + T_N + F_P + F_N} \qquad (8.1)$$

where, $T_P$ = true positive, $T_N$ = true negative, $F_P$ = false positive, and $F_N$ = false negative.

### 8.3.1  Is Accuracy Sufficient to Justify the Performance?

Accuracy measure performs well when there is a balance of samples among each class. When there is an imbalance of input samples among classes, it is not possible to justify whether a model is performing well or not. We can look at Fig. 8.2a as an example of this case. This figure summarizes that the used model can perfectly predict all of the 998 walk data, whereas it could predict 1 out of 2 run data correctly. The accuracy of this model is very high (99.9%), which does not correctly sum up the performance of this model as we can not represent the weakness of this model in terms of classifying run data showing this overall accuracy.

**Fig. 8.2** Accuracy as a performance measure for classification problem

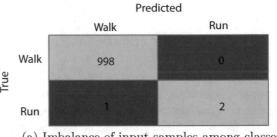

(a) Imbalance of input samples among classes

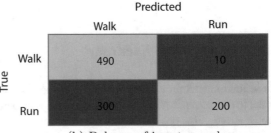

(b) Balance of input samples.

If we make a balance of samples between input classes and use the same model, we will find from Fig. 8.2b that the accuracy will be dropped down to 69.0% only. This is because the model can correctly classify walk but fails to recognize run data. This limitation of this model was not present during the previous case because there were very few amounts of run data, which was the reason for the overall higher accuracy of the model with a false sense. When the cost of misclassification of the minor class samples is very high, this could be a real problem.

## 8.4 Precision or Positive Predictive Value

Precision is also known as the positive predictive value. It is defined as the proportion of predicted positives that are positive. It means the percentage of results, which are relevant. In a word, we can say that precision represents how many decided labels are appropriate. We can define precision by the following formula,

$$Precision = \frac{T_P}{T_P + F_P} \tag{8.2}$$

where, $T_P$ = true positive and $F_P$ = false positive.

## 8.5 Recall or Sensitivity or True Positive Rate

Recall is also known as sensitivity or true positive rate. It is defined as the proportion of actual positives that are predicted positive. It means the percentage of total relevant results that have been correctly classified by the model. In a word, we can say that recall represents how many relevant items are selected. We can define recall by the following formula,

$$Recall = \frac{T_P}{T_P + F_N} \tag{8.3}$$

where, $T_P$ = true positive and $F_N$ = false negative.

## 8.6 Precision Versus Recall

There lies a trade-off between precision and recall. If we need to maintain a good recall percentage, it will let the model keep generating results which may not be accurate. This will lower the precision. On the other hand, if the model is bound to predict the result with high precision rate, it can not keep generating non-accurate results, which will lower the recall. There is no possible way to maximize both of these metrics at the same time. There can be generalized problems where we may choose to give priority to either precision or recall value but in most of the cases, there is a metric named F1 score that takes into account both precision and recall. We can design our model to maximize this metric to make the model better.

## 8.7 F1 Score

F1 score is defined as the harmonic mean between precision and recall. This metric takes both false positives and false negatives into account. For uneven class distribution, the F1 score is a very good measure. We can define the F1 score by the following expression,

$$F1\ score = 2 \times \frac{Precision \times Recall}{Precision + Recall} \tag{8.4}$$

## 8.8   $F_\beta$ Score

This metric measures the effectiveness of retrieval concerning a user who attaches $\beta$ times as much importance to recall as precision. This is the weighted harmonic mean between precision and recall, which can be defined by the following formula,

$$F_\beta = \frac{(1 + \beta)^2 \times T_P}{(1 + \beta)^2 \times T_P + \beta^2 \times F_N + F_P} \tag{8.5}$$

where, $\beta$ is the shape parameter.

## 8.9   Specificity or True Negative Rate

Specificity is also known as the true negative rate. It is defined as the proportion of actual negatives that are predicted negative. We can define specificity by the following equation,

$$Specificity = \frac{T_N}{T_N + F_P} \tag{8.6}$$

## 8.10   Positive Likelihood

This is a likelihood that a predicted positive is an actual positive. This can be defined by the following formula,

$$Positive\ likelihood = \frac{Sensitivity}{1 - Specificity} \qquad (8.7)$$

## 8.11  Negative Likelihood

This is a likelihood that a predicted negative is an actual negative. This can be defined by the following formula,

$$Negative\ likelihood = \frac{1 - Sensitivity}{Specificity} \qquad (8.8)$$

## 8.12  Balanced Classification Rate (BCR)

Balanced Classification Rate (BCR) is also known as balanced accuracy. This metric combines the sensitivity and specificity metrics and it can be defined by the following equation,

$$
\begin{aligned}
BCR &= \frac{1}{2}(True\ Positive\ Rate + True\ Negative\ Rate) \\
&= \frac{1}{2}(\frac{T_P}{T_P + F_N} + \frac{T_N}{T_N + F_P})
\end{aligned} \qquad (8.9)
$$

## 8.13  Balanced Error Rate (BER)

This metric is also known as Half Total Error Rate (HTER) and it can be defined by the following equation,

$$BER = 1 - BCR \qquad (8.10)$$

## 8.14  Youden's Index

This is defined as follows with the values of sensitivity and specificity,

$$Youden's\ Index\ (J) = Sensitivity + Specificity - 1 \qquad (8.11)$$

## 8.15 Matthews Correlation Coefficient

This is a measure of the quality of classification problems. This metric takes into account the false positive and false negative and it can be used for classes with different sizes too. This reflects a correlation coefficient between the binary classifications observed and expected, and returns a value between $-1$ to $+1$. A coefficient of $+1$ denotes an accurate prediction, 0 denotes no better than random prediction and $-1$ indicates total disagreement between prediction and observation. This is also known as the *phi coefficient*. We can represent the Matthews Correlation Coefficient (MCC) using the following expression,

$$MCC = \frac{(T_P \times T_N) - (F_P \times F_N)}{\sqrt{(T_P + F_P) + (T_P + F_N) + (T_N + F_P) + (T_N + F_N)}} \qquad (8.12)$$

where, $T_P$ = true positive, $T_N$ = true negative, $F_P$ = false positive, and $F_N$ = false negative.

## 8.16 Discriminant Power Normalized Likelihood Index

The Discriminant Power Normalized Likelihood Index ($D_P$) is another performance measure for classifiers, which is defined by the following equation,

$$D_P = \frac{\sqrt{3}}{\pi}[log(\frac{Sensitivity}{1 - Specificity}) + log(\frac{Specificity}{1 - Sensitivity})] \qquad (8.13)$$

## 8.17 Cohen's *kappa* Coefficient ($\kappa$)

This coefficient measures the performance of a classifier on the basis that how well the classifier performed in comparison to the case where it would have performed simply by chance. We can say that a model will have a high *kappa* ($\kappa$) score if there is a big difference between the accuracy and the null error rate. Null error rate defines the chances of the wrong prediction in case of always predicting the majority class.

## 8.18 Confusion Matrix

This is one of the most-widely explored representations in the field of classification and related areas. This is a table-type representation of the performance of a classifier on a set of test data for which ground truths are known. This table gives a summary

**Fig. 8.3** Typical representation of a confusion matrix

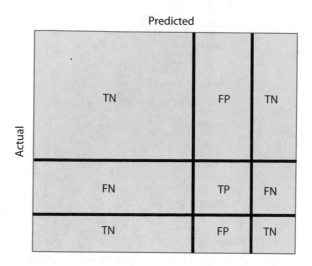

of the prediction results of a classification problem. A total number of correct and incorrect predictions are reviewed with count values and broken down by each class. It shows when the model gets confused while predicting the results. Confusion matrix not only specifies the errors made by a classifier but also gives an insight into the type of the errors being made. A typical representation of a confusion matrix has been shown in Fig. 8.3, where four parameters are associated namely true positive, true negative, false positive, and false negative that have been described earlier.

The term *confusion* in a confusion matrix or confusion table, determines the classes that are confused or misclassified by other classes. Figure 8.4 demonstrates one example where 6 activity classes are mentioned. Top row's 98.4% value provides the recognition result or accuracy for 'stay' class. However, this 'stay' class is confused with 'walk' class by 0.7% and 'jog' class by 0.9%. The values in the 1st row summed up as 100%. This is one way to represent confusion matrix. From this Table, we can predict that 'upstairs' class is highly confused or misclassified by 'walk' activity class (6.9%). Similarly, we can notice that the 'walk' class is also confused with 'upstairs' class (3.5%). So, both 'walk' and 'upstairs' classes are getting mixed up or messed up or confused by each other. So, if any kind of preprocessing or improvement is possible to make that can differentiate these two confused activity classes, we can improve the overall accuracy.

Through the confusion matrix, we can discuss and analyze the reasons for misclassification or, try to decipher the rationality of any error in accuracy per class. Using different programming languages or visualization tools, we can demonstrate confusion matrix by varied combinations of colors, instead of the numbers. The diagonal values (in *black* squares) are the accuracy rates for the classes.

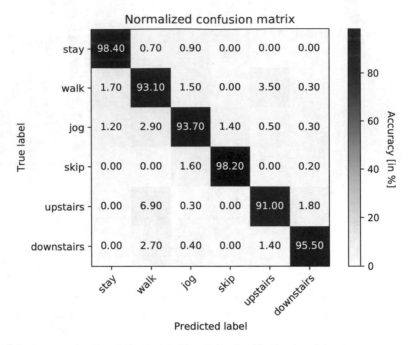

**Fig. 8.4** An example of confusion matrix for activity classification in a dataset

## 8.19   Receiver Operating Characteristics (ROC) Curve

Receiver Operating Characteristics (ROC) curve is one of the most important evaluation metrics, which is a probability curve that measures the performance of classification problems under various threshold settings. Area Under the Curve (AUC) term is related to the ROC curve, which represents the degree of measure of separability [15]. If the AUC is higher, it means that the model is good at separating the classes accurately. The ROC curve is plotted with the true positive rate (TPR) or recall or sensitivity against the false positive rate (FPR) or specificity, where TPR is on the y-axis and FPR is on the x-axis. Recommendations for using the ROC curve can be found in [17].

A basic figure of ROC curve has been shown in Fig. 8.5, where we can see that an excellent model has AUC near to the 1 with a good measure of separability. On the other hand, AUC near 0 represents a poor model with worst separability measure. If a model has AUC of 0.5, we can say that the model will predict randomly.

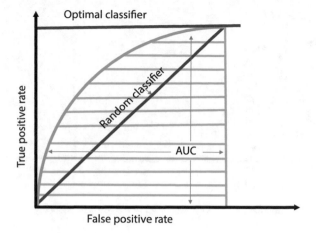

**Fig. 8.5** Basic representation of ROC curve

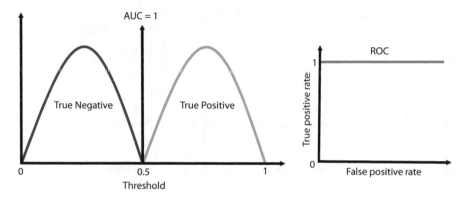

**Fig. 8.6** Ideal probability distribution of ROC with AUC = 1

### 8.19.1  Probability Distribution of ROC Curve

We can see the distributions of probabilities from ROC probability distribution curves as shown in Figs. 8.6, 8.7, 8.8, and 8.9. We can see an ideal situation in Fig. 8.6, where the distribution curves do not overlap. This means that the model has an ideal measure of separability and the model can separate between the positive and negative classes accurately.

The type-1 and type-2 errors occur when the two distribution curves overlap, as shown in Fig. 8.7. AUC value drops in these cases. AUC value of 0.6 means that there is a 60% chance that the model can perfectly separate between the positive and negative classes.

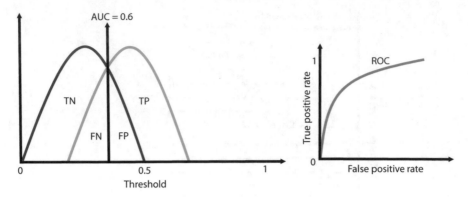

**Fig. 8.7**   Probability distribution of ROC with AUC = 0.6 (drop of accuracy)

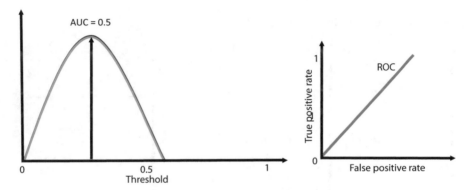

**Fig. 8.8**   Probability distribution of ROC with AUC = 0.5 (random prediction)

The worst-case has been shown in Fig. 8.8, where AUC is 0.5. In this case, the model will predict randomly with no discrimination capacity to predict between positive and negative class.

AUC value of 0 has been shown in Fig. 8.9. In this case, the model will reciprocate the classes. For example, it will predict the positive classes as negative and negative classes as positive.

### 8.19.2   ROC Curve for Multiclass Model

In case of a multiclass model with $n$ number of classes, we can plot $n$ number of ROC curves for $n$ number of classes using One versus All strategy. As an example, for a classification problem with three classes ($a$, $b$, and $c$), we can plot one ROC for $a$, classified against $b$ and $c$; another for $b$, classified against $a$ and $c$; and the last one for $c$, classified against $a$ and $b$.

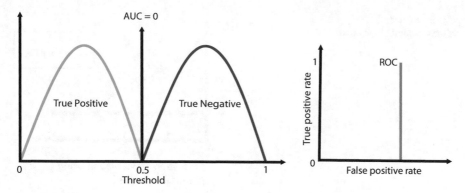

**Fig. 8.9** Probability distribution of ROC with AUC = 0 (reciprocal prediction)

**Fig. 8.10** A basic representation of cumulative gains chart

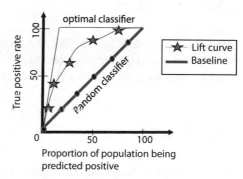

## 8.20 Cumulative Gains and Lift Charts

To measure the model performance, two well-known graphical aids are Cumulative Gains and Lift Charts. Lift measures how much effective a predictive model is. We can calculate the lift using the ratio between the results obtained with and without the predictive model. A lift curve and a baseline are presented in both of the curves. The larger the area between the lift curve and the baseline, the better the model.

### 8.20.1 Cumulative Gains Chart

This is a plot of the true positive rate as a function of the proportion of the population is predicted positive, controlled by some classifier parameters or thresholds. True positive rate is plotted on the y-axis and proportion of the population being predicted positive on the x-axis. The baseline defines the overall response rate. Using the predictions of the response model, the lift curve is drawn mapping the true positive rate points. This chart is shown in Fig. 8.10.

**Fig. 8.11**  A basic
representation of lift chart

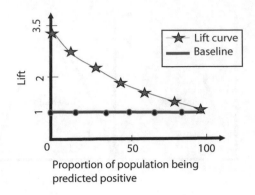

## 8.20.2  *Lift Chart*

The actual lift is shown by this chart. To plot the chart, we need to calculate the points on the lift curve by determining the ratio between the result predicted by the model and the result using no model. This chart is shown in Fig. 8.11.

## 8.21  Conclusion

This chapter provides a basic summary concerning the evaluation matrices namely accuracy, precision, recall, F1 score, confusion matrix, etc. to justify the performance of classification models. We have also shown the importance of Receiver Operating Characteristics (ROC) curve with the probability distribution for the different Area Under the Curve (AUC). We have also analyzed some common confusions about the trade-off between precision and recall in this chapter.

## 8.22  Think Further

1. Why proper evaluation of performance is necessary in human activity recognition research?
2. What are the four essential concepts of classification?
3. Differentiate among true positive, true negative, false positive, and false negative.
4. When classification accuracy can not properly measure the performance?
5. What are the common evaluation matrices?
6. What is the trade-off between precision and recall?
7. Mention the importance of confusion matrix.
8. What is AUC in ROC curve?
9. Specify the probability distributions of ROC curve.

10. What are the importance of Cumulative Gain and Lift Charts?
11. Read 10 good papers from different journals on human activity recognition, gait recognition, image classification, medical imaging classification, or related. Then find the matrices those papers explored. Try to understand these issues in different fields.
12. What are the missing matrices or points you think, in terms of evaluating activity classification?

# References

1. Antar, A.D. Ahad, M.A.R., Shahid, O.: Vision-based action understanding for assistive health-care: a short review. In: IEEE CVPR workshop (2019)
2. Ahad, M.A.R.: Vision and sensor based human activity recognition: Challenges ahead (2020)
3. Antar, A.D., Ahmed, M., Ahad, M.A.R.: Challenges in sensor-based human activity recognition and a comparative analysis of benchmark datasets: a review. In: 2019 Joint 8th International Conference on Informatics, Electronics & Vision (ICIEV) and 2019 3rd International Conference on Imaging, Vision & Pattern Recognition (icIVPR), pp. 134–139. IEEE (2019)
4. Ahad, M.A.R.: Motion History Images for Action Recognition and Understanding. Springer Science & Business Media, Berlin (2012)
5. Ahad, M.A.R.: Computer Vision and Action Recognition: A Guide for Image Processing and Computer Vision Community for Action Understanding, vol. 5. Springer Science & Business Media, Berlin (2011)
6. Hossain, T., Islam, M.S., Ahad, M.A.R., Inoue, S.: Human activity recognition using ear-able device. In: Proceedings of the 2019 ACM International Joint Conference on Pervasive and Ubiquitous Computing and Proceedings of the 2019 ACM International Symposium on Wearable Computers, pp. 81–84. ACM (2019)
7. Tazin, T., Hossain, T., Ahad, M.A.R., Inoue, S.: Activity recognition by using lorawan sensor. In: 2018 ACM International Joint Conference on Pervasive and Ubiquitous Computing and the 2018 International Symposium on Wearable Computers (UbiComp/ISWC) (2018)
8. Ahmed, M., Antar, A.D., Ahad, M.A.R.: An approach to classify human activities in real-time from smartphone sensor data. In: 2019 Joint 8th International Conference on Informatics, Electronics Vision (ICIEV) and 2019 3rd International Conference on Imaging, Vision Pattern Recognition (icIVPR), pp. 140–145 (2019)
9. Sokolova, M., Lapalme, G.: A systematic analysis of performance measures for classification tasks. Inf. Process. Manag. **45**(4), 427–437 (2009)
10. Demšar, J.: Statistical comparisons of classifiers over multiple data sets. J. Mach. Learn. Res. **7**(Jan), 1–30 (2006)
11. Powers, D.M.: Evaluation: from precision, recall and f-measure to roc, informedness, marked-ness and correlation (2011)
12. Saito, T., Rehmsmeier, M.: The precision-recall plot is more informative than the roc plot when evaluating binary classifiers on imbalanced datasets. PloS one **10**(3) (2015)
13. Saktheeswari, M., Balasubramanian, T.: Performance analysis of classifier models to predict thyroid disease (2018)
14. Fawcett, T.: Introduction to receiver operator curves. Pattern Recognit. Lett. **27**, 861–874 (2006)
15. Hand, D.J., Till, R.J.: A simple generalisation of the area under the roc curve for multiple class classification problems. Mach. Learn. **45**(2), 171–186 (2001)
16. Provost, F., Fawcett, T.: Robust classification for imprecise environments. Mach. Learn. **42**(3), 203–231 (2001)
17. Robert Gilmore Pontius and Benoit Parmentier: Recommendations for using the relative operating characteristic (roc). Landsc. Ecol. **29**(3), 367–382 (2014)

# Chapter 9
# Deep Learning for Sensor-Based Activity Recognition: Recent Trends

**Abstract** The field of human activity recognition (HAR) using different sensor modalities poses numerous challenges to the researchers working in this domain. Though traditional pattern recognition approaches performed well in this regard earlier, the cost of poor generalization and the cost of shallow learning due to the handcrafted features have opened a new door for deep learning in this field. This chapter discusses the importance of deep learning in sensor-based activity recognition explaining the deep models and their use in previous research works. This chapter also represents the importance of transfer learning and active learning in this field, that are new research topics. Finally, this chapter shows the challenges of using deep models along with feasible solutions.

## 9.1 Evaluation of Deep Learning in Sensor-Based HAR

Sensor-based Human Activity Recognition (HAR) has been explored by many research communities and industries for various applications—along with various challenges ahead to deal with [1–8]. Because of the complexities related to activity recognition, choosing sensor modality, and multiple numbers of subjects, sensor-based human activity recognition (HAR) is one of the most challenging domains for the researchers. Investigation of proper machine learning techniques is necessary as they are efficient in the case of extracting and learning knowledge from raw sensor data. The research in the area of HAR has begun considering this a conventional pattern recognition problem [9]. Typical pattern recognition approaches and general machine learning algorithms namely Support Vector Machine, Naive Bayes, Decision Tree, Hidden Markov Model, etc. have made tremendous progress in the field of sensor-based HAR. These algorithms performed well with satisfactory results

- In the case of a controlled environment with fewer amount of labeled data, and
- In the case of specific domain knowledge requirement.

These traditional methods follow the concept of shallow learning, which is heavily dependent on feature engineering from the data. These methods depend heavily on

© Springer Nature Switzerland AG 2021

M. A. R. Ahad et al., *IoT Sensor-Based Activity Recognition*, Intelligent Systems Reference Library 173, https://doi.org/10.1007/978-3-030-51379-5_9

hand-crafted heuristic feature extraction and are restricted by awareness of the human domain [10]. This learning method incorporates the learning of shallow features only, which performs poor in the case of unsupervised and incremental tasks. Due to these restrictions, the performance of conventional pattern recognition approaches is limited in terms of accuracy and generalization of the model.

## 9.2  Why Deep Learning Can Increase the Performance of HAR?

The earlier practice of utilizing conventional pattern recognition approaches and general classification techniques for recognizing the daily human activities from sensor data has numerous limitations. These methods restrict the model to generate for one domain to extend into another domain. This domain transfer is harder for these techniques. In spite of the progress of conventional algorithms, the major drawbacks are discussed in this section.

Firstly, the conventional models are dependent on hand-crafted features dependent on human-experience and specific domain knowledge. In the case of task-specific environmental setup, these models may perform well; but for more generalized case in real-time, the performance is poor and it will demand more running time to develop a successful activity recognition model.

Secondly, the features learned in the case of hand-designed approaches are shallow in nature [11] and basically, they represent some statistical information like mean, standard deviation, variance, amplitude, frequency, and so on [12–14]. Though this is easy to recognize low-level activities like jogging, running, walking, etc. using these shallow features, it will be difficult or near impossible to detect complex activities that are similar. Besides, it is also very difficult to capture complex movements for high level and context-aware activities that involves a series of several microcavities using shallow learning [15, 15, 16]. For example, it will be harder and nearly impossible to recognize and differentiate between the activities "drinking water" and "drinking milk", using conventional statistical feature-based models.

Thirdly, conventional models show poor performance in the case of unsupervised learning tasks [10] because it requires a large amount of labeled data to train these models and this task is difficult [17]. In the real case, most of the activity data are unlabeled, and labeled data are always incorporated with human labeling error, which is another reason for the poor performance. Deep neural networks, on the other hand, can easily exploit unlabeled data to train the model.

A further downside of current machine learning models is that they depend on learning from static data, whereas, in the real-time, all activity data comes in a stream. This stream of data requires strong and incremental learning, which is not possible using conventional pattern recognition approach.

Deep learning models have overcome some of these limitations imposed by the volatile, chaotic, and complex nature of the activity data, as they can hierarchically

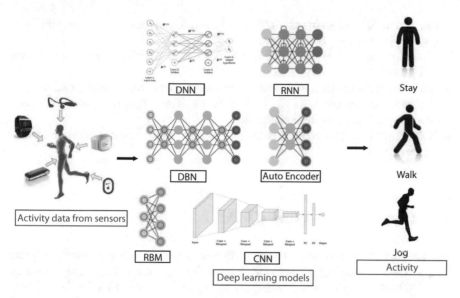

**Fig. 9.1** A basic block diagram of sensor-based human activity recognition using deep learning methods

learn the features directly from the data. These self-evolving feature-based deep learning models can learn the features automatically, which releases the pressure of designing the features manually.

Besides, the use of deep neural networks can extract high-level features in the deep layers, which can overcome the problem of complex activity recognition, and multiple user problem. Moreover, it is also possible to redeploy the domain transfer of the activity models. It is also possible to exploit the huge amount of unlabeled data using deep generative models [18]. Moreover, we can also transfer the knowledge of trained deep models into labeled data to new tasks containing few or no labels. In the Fig. 9.1, we have shown a general pipeline of utilizing deep learning networks for the sensor-based human activity recognition task.

## 9.3   Deep Learning Models for Human Activity Recognition

In this section, we have given a brief description of the deep learning models that are generally used for the sensor-based human activity recognition task. In general, following deep learning models have been mostly used in previous HAR researches with modifications and updates [9],

- **Deep Neural Network (DNN):** Deep fully-connected neural network, artificial neural network (ANN) with deep layers.

- **Recurrent Neural Network (RNN):** A network with time correlation and Long Short-Term Memory (LSTM) layers.
- **Convolutional Neural Network (CNN):** A network with multiple convolutions and pooling layers.
- **Deep Belief Network (DBN) and Restricted Boltzmann Machine (RBM).**
- **Stacked Autoencoder (SAE):** This network learns feature by the decoding-encoding technique of the autoencoder.
- **Hybrid Models:** This network is designed using the combination of different deep learning models.

## 9.4  Deep Neural Network (DNN)

The concept of deep learning begins with Perceptron, which is basically a linear binary classifier with a single layer. Multilayer Perceptron is called neural networks. The depth of neural networks is decided using the number of hidden layers. Traditional neural networks with very few hidden layers are called Artificial Neural Network (ANN). It is possible to develop Deep Neural Networks (DNN) from ANN using denser hidden layers. Because of the presence of dense hidden layers, DNN can learn from a large amount of data automatically without the need of hand-crafted features. It is also possible to use DNN as dense layers for other deep models. A basic block diagram of a DNN is shown in Fig. 9.2.

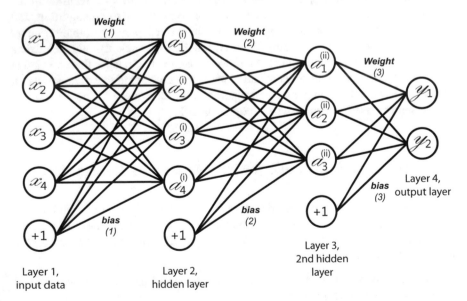

**Fig. 9.2**  A basic block diagram of Deep Neural Network (DNN)

We will now discuss some previous research works using DNN. Research work [19] built a Deep Neural Network model and fed hand-crafted features from sensor data into it. Research work [20], on the other hand, performed Principal Component Analysis (PCA) before applying DNN model. We can say that in these works, DNN has been used only as a classification model after extraction of hand-engineered features. These methods showed poor performance in terms of generalization and accuracy being shallow.

Automatic feature learning and classification can increase the generalization capability of a model. This task has been done in the research [21], where they have utilized DNN with multiple hidden layers and performed well in the case of multi-dimensional and complex HAR data. More hidden layers increase the generalization capability of a model and the model can train well even for the complex activity data [10].

## 9.5   Convolutional Neural Network (CNN)

The primary advantage of using Convolutional Neural Network (CNN) in the case of human activity recognition is the identification capability of notable patterns in the sensor data. CNN is designed using convolution layers followed by pooling and fully-connected layers as shown in Fig. 9.3. Besides showing promising performance in the case of text analysis, image classification, and speech recognition, CNN has also advantages in the case of sensor data for activity recognition. The reasons behind this are the local dependency of sensory data as nearby sensor signals in the case of activity data are highly correlated. Besides, there is a variation of scale in terms of frequency for the performed activities at various speeds. In the lower layers of CNN, the basic characteristics of each basic movement of human activities are obtained, whereas, the higher layers can obtain the important patterns of a combination of multiple basic movements. This is the reason behind the utilization of Convolutional Neural Networks in a number of the previous researches related to HAR.

A multilayer CNN can have multiple convolutions and pooling operators in each layer that can learn multiple salient patterns from signals, which are jointly considered. It is possible to obtain translation invariance when these operators are applied

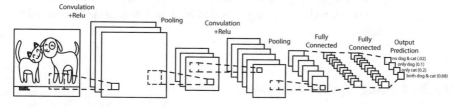

**Fig. 9.3** A basic block diagram of Convolutional Neural Network (CNN). Here, 'ReLu' is the Rectified Linear Unit. It is one of the widely-used activation functions

to the local signals [22–24]. Because of the presence of multiple channel time-series signals in HAR, it is not possible to apply traditional CNN directly. The CNN should be designed in such a way that it can be applied along temporal dimension with sharing units among multiple sensors. The following things must be considered while designing the CNN architecture for HAR.

### 9.5.1 Adaptation of Input Signal

CNN architectures are designed for images as input, whereas sensor signals are generally multi-dimensional time-series signals. The primary task is to adapt the raw sensor data in such a way that it forms the structure of a virtual image. This adaptation can be done in one of the following ways [9],

- Data-driven approach,
- Model-driven approach.

In the data-driven approach, each dimension of the multi-dimensional data is considered as a single channel, and one-dimensional convolution is performed. After performing convolutions and pooling, the outputs of each channel are flattened. In one of the earliest research works by [25], each dimension of accelerometer sensor data had been treated like a single RGB channel of an image followed by convolution and pooling separately. Research works [11] utilized one-dimensional convolution in the same temporal window. This method permits weight sharing in the case of multi-sensor CNN. Research work [21, 26–28] focused on resizing the kernel of convolution layer in order to achieve the best kernel for sensor data. In short, we can say that in the data-driven approach, 1D sensor data is considered as a 1D image. However, this approach ignores the dependencies between dimension and sensors, which can turn into poor performance.

Another approach of input data adaptation is called the model-driven approach. In this method, the input sensor data is resized into a two-dimensional image so that 2D convolution can be applied. Research work [29] formed an image combining sensor data from all dimensions. On the other hand, research work [30] transformed time series sensor data into an image by designing a complex algorithm. Research work [31] showed an example of modality transformation by converting pressure sensor data into an image. Temporal correlation of sensor data can be utilized in this approach, which is a benefit but it requires domain knowledge to map time series data into an image.

### 9.5.2 Tuning Hyperparameters

There are some hyperparameters for a CNN architecture that we can play with for better performance. These are

- The number of features and size of features for convolution layer,
- Window size and window stride for pooling layer, and
- The number of neurons for the fully connected layer.

Choosing the type of pooling after performing convolution is another important task. Most of the research works including [27, 29, 32] performed max or average pooling. Pooling operation helps in two ways: speeding up the training process and limiting overfitting.

### 9.5.3  Weight Sharing

To speed up the training procedure, research works [28, 33] introduced weight sharing technique, whereas, research work [25] proposed partial weight sharing because of the variation of sensor data in different cases. It also showed that this partial weight sharing technique improves CNN performance.

## 9.6  Recurrent Neural Network (RNN)

Recurrent Neural Network (RNN) gives us the benefit to feed a series of input without any predetermined size. RNN can catch the connections among inputs meaningfully. Because of the utilization of the temporal correlations among neurons, RNN are widely used for natural language processing (NLP) and speech recognition. Another benefit of using RNN is the capability of learning from the past, whereas, basic feedforward neural networks can remember only what they have learned during the training phase.

A general RNN has short-term memory along with the problem of vanishing gradients. Vanishing gradient means that the gradient values are so small that the model stops learning. The extension of RNN with Long Short-Term Memory (LSTM) can extend the memory of the network. By using LSTM networks, RNN can remember the inputs over a long period. It also keeps the gradient steep enough to eliminate the vanishing gradient problem. RNN architecture and RNN with LSTM have been shown in Figs. 9.4 and 9.5.

Most of the human activity recognition research works focused on computational cost, resource consumption, and training speed while using RNN [21, 34–37]. Research work [36] has proposed a good model with high throughput after the investigation of numerous model parameters. A Binarized Long Short-Term Memory Network (B-BLSTM-RNN) model has been proposed by the research work [34], where the inputs, outputs, and weight parameters are binary values. The B-BLSTM-RNN is based on the bidirectional Long Short-Term Memory Recurrent Neural Network (BLSTM-RNN). The overall performance of RNN is promising in the case of HAR with more research concerns about the computational cost and resource limitations.

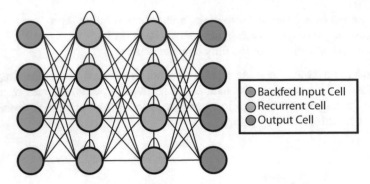

**Fig. 9.4**  A basic block diagram of Recurrent Neural Network (RNN)

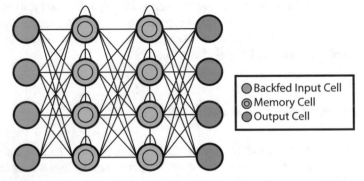

**Fig. 9.5**  A basic block diagram of Recurrent Neural Network (RNN) with Long Short-Term Memory (LSTM)

## 9.7   Stacked Autoencoder (SAE)

The latent representations of the input values are learned through the hidden layers in the case of an autoencoder (AE). This procedure can seem as an encoding-decoding procedure. The basic advantage of the autoencoder is the learning capability of advanced features through an unsupervised learning scheme. The stacking of some autoencoders is generally called Stacked Autoencoder (SAE), where every layer is treated as the basic model. The learned features are stacked with labels so that they can be used as classifiers after several rounds of training. Figure 9.6 shows a basic structure of an autoencoder.

Research works [38, 39] utilized the stacked autoencoder (SAE) for human activity recognition with greedy layer-wise pre-training technique followed by fine-tuning. On the other hand, research work [40] added KL divergence and noise to the cost function and this method improved the performance. The unsupervised feature learning capability of SAE has made it a powerful tool with the cost of optimal solution due to the too much dependency on layers and activation function.

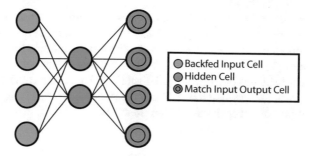

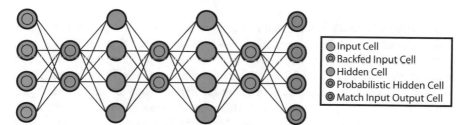

**Fig. 9.6** A basic block diagram of an autoencoder (AE)

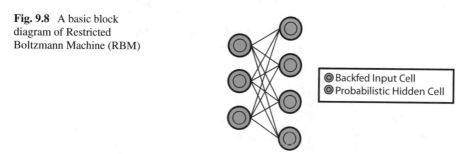

**Fig. 9.7** A basic block diagram of Deep Belief Network (DBN)

**Fig. 9.8** A basic block diagram of Restricted Boltzmann Machine (RBM)

## 9.8 Deep Belief Network (DBN) and Restricted Boltzmann Machine (RBM)

Restricted Boltzmann Machine (RBM) is an energy-based probabilistic model consisting of visible variables and hidden variables. This can be seen as a fully-connected, undirected graph consisting of visible and hidden layers [18]. If the RBM's are stacked, it is called Deep Belief Network (DBN) by treating two consecutive layers as an RBM, where both DBN and RBM are followed by fully-connected layers. The architectures of DBN and RBM have been shown in Figs. 9.7 and 9.8.

In case of the RBM-based researches [41–43], most of the works concerned Gaussian RBM in the first layer and binary RBM for the rest of the layers during pretraining. Research work [44] designed a multi-modal RBM for HAR with multi-modal sensors, and each RBM was created for each sensory modality. Research work [45], on the other hand, employed pooling action after fully-connected layer for the extraction of more important features. Contrastive gradient (CG) method has been utilized by research work [46] to update the weight in fine-tuning. The network can search and converge easily by this procedure in all directions. Research work [47] employed offline training by implementing RBM on mobile phones, as RBM is lightweight and it has the capability of unsupervised feature learning.

## 9.9   Hybrid Models

When some deep models are combined to enhance the overall performance, it is called a hybrid model. Research work [47–49] combined CNN and RNN, showing that the combination of CNN and recurrent dense layer performs better than CNN with general dense layers [31, 50]. When the ability of CNN to capture spatial relationship and RNN to capture the temporal relationships are combined, it can help to recognize complex activities with different time span and signal distributions. There are some other research works where hybrid models have been generated using CNN + SAE [51] and CNN + RBM [52]. In these works, CNN has been utilized for the feature extraction tasks, and generative models speed up the training procedure. The researchers are focused on hybrid models nowadays where performance is a major concern. We have shown a basic block diagram of a hybrid model using DNN and CNN based on the research work [53] in Fig. 9.9.

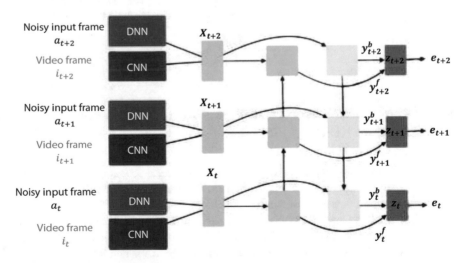

**Fig. 9.9**   A basic block diagram of hybrid model using DNN and CNN

## 9.10 Choosing the Best Deep Model for Sensor-Based HAR

The existence of complex activities in sensor-based HAR is one of the most challenging fields of research. Though deep models have solved the problems of hand-crafted features and shallow learning, it is required to choose the best model and optimized hyperparameters for better performance. Research work [21] performed around four thousand experiments using different setups on DNN, CNN, and RNN on some public datasets. Based on this work, RNN with LSTM demonstrated good performance for recognizing short activities with the natural order, whereas CNN showed good performance for the long term and repetitive activities. Because RNN captures the temporal relationships between sensor readings and CNN can learn deep features containing recursive patterns, it is better to use CNN for multimodal sensor signals. The features can be integrated through multichannel convolutions [25, 29, 54]. In the case of CNN, data-driven approach is better than the model-driven approach. The reason behind this is the capability of transforming sensor data into the virtual image.

In the case of RBM and autoencoder, it is required pre-training before fine-tuning. Multilayer RBM and SAE can amplify the performance than single layer RBM and SAE. We can state that there is no perfect model, which can overcome all the problems in every situation. It is required to choose the model based on the scenario, computational cost, resource utilization and limitation, and performance. In Table 9.1, we have summarized a number of previous works on sensor-based human activity recognition using deep models.

**Table 9.1** Summary of several previous research works on sensor-based human activity recognition utilizing deep learning techniques

| Reference | Method | Dataset |
|-----------|--------|---------|
| [21] | DNN, CNN, and RNN (RNN performance was better than CNN in the case of short-term activity) | OPPORTUNITY [55], PAMAP2 [54], and Daphnet Gait [21] |
| [50] | A deep network comprising of convolutional and LSTM layers | OPPORTUNITY [55] and Skoda [56] |
| [57] | Multilayer CNN model with alternating convolutional and pooling layers | Dataset owned by [57] with 30 subjects |
| [58] | Short-term Fourier transform of the accelerometer data as an input to the proposed CNN network | Skoda [56], WISDM [59], Daphnet Gait [21], and ActiveMiles [58] |
| [60] | Restricted Boltzmann Machine (RBM)-based model without hardware constraints | OPPORTUNITY [55], Transportation and Physical [60], and Indoor/Outdoor [44] |
| [61] | Ensembles of deep LSTM networks using wearable sensing data | Dataset owned by [61] |

(continued)

**Table 9.1**  (continued)

| Reference | Method | Dataset |
|---|---|---|
| [62] | Hand-crafted features and the CNN-derived features—fed to k-Nearest Neighbor (kNN) | Dataset owned by [62] (data from wrist and waist) |
| [63] | Multichannel CNN architecture for multiple sensor data | Dataset owned by [63] |
| [38] | Stack Autoencoder (SAE) | UCI Smartphone [38] |
| [59] | Restricted Boltzmann machine (RBM) | Skoda [56], WISDM [59], and Daphnet Gait [21] |
| [26] | CNN | Dataset owned by [26] |
| [64] | CNN | WISDM [59] |
| [65] | DNN | Dataset owned by [65] |
| [34] | RNN | OPPORTUNITY [55], PAMAP2 [54], and dataset owned by [34] |
| [46] | DBN | Dataset owned by [46] |
| [66] | CNN | Dataset owned by [66], and OPPORTUNITY [55] |
| [35] | RNN | OPPORTUNITY [55], Skoda [56], and PAMAP2 [54] |
| [29] | CNN | Skoda [56] and MHEALTH [67] |
| [67] | CNN | MHEALTH [67] |
| [41] | RBM | Dataset owned by [41] |
| [68] | CNN | Dataset owned by [68] |
| [69] | RBM | HASC [69] |
| [36] | RNN | HASC [69] |
| [30] | CNN | UCI Smartphone [38], USC-HAD [30], and SHO [30] |
| [70] | CNN | Dataset owned by [70] |
| [32] | CNN | Dataset owned by [32] |
| [71] | CNN | Dataset owned by [71] |
| [72] | RBM | Dataset owned by [72] |
| [42] | RBM | Dataset owned by [42] |
| [73] | CNN | Dataset owned by [73] |
| [45] | RBM | Dataset owned by [45] |
| [40] | SAE | UCI Smartphone [38] |
| [52] | CNN, RBM | Dataset owned by [52] |
| [74] | CNN | Dataset owned by [74] |
| [75] | CNN | OPPORTUNITY [55] and Skoda [56] |
| [76] | RNN | OPPORTUNITY [55], Skoda [56], Ambient kitchen [43], USC-HAD [30], and Daphnet Gait [21] |

(continued)

**Table 9.1** (continued)

| Reference | Method | Dataset |
|---|---|---|
| [43] | RBM | OPPORTUNITY [55], Skoda [56], Ambient kitchen [43], and Daphnet Gait [21] |
| [27] | CNN | PAF [27] |
| [44] | RBM | Heterogeneous [48] |
| [77] | CNN | UCI Smartphone [38] |
| [28] | CNN, RNN, and DNN | Dataset owned by [28] |
| [31] | CNN, RNN | Dataset owned by [31] |
| [19] | DNN | Dataset owned by [19] |
| [20] | DNN | UCI Smartphone [38] |
| [78] | CNN | Dataset owned by [78] |
| [39] | SAE | Dataset owned by [39] |
| [11] | CNN | OPPORTUNITY [55] and Ambient kitchen [43] |
| [48] | CNN, RNN | Dataset owned by [48] and Heterogeneous [48] |
| [33] | CNN | Dataset owned by [33] |
| [79] | CNN | OPPORTUNITY [55], Skoda [56], and Actitracker [25] |
| [80] | DNN | Dataset owned by [80] |
| [47] | RBM | Dataset owned by [47] |
| [81] | DBN | OPPORTUNITY [55], USC-HAD [30], and DSADS [81] |
| [82] | CNN | Dataset owned by [82] |
| [83] | DNN | Dataset owned by [83] |
| [51] | CNN, SAE | PAMAP2 [54] |
| [54] | CNN | PAMAP2 [54] and Daphnet Gait [21] |
| [84] | DBN | UCI HAPT [85] |
| [86] | Three parallel CNN | UCI HAR [87] and WISDM [59] |
| [88] | Ensemble Network with RNN | SHL [89] |
| [90] | Shallow Neural Network | SHL [89] |
| [91], [92] | CNN, RNN with LSTM, | OU-ISIR Gait [93] |

## 9.11 Transfer Learning in Sensor-Based HAR

Because of the high computational cost and training time related to the processing of large HAR datasets, researchers have focused on transferring the knowledge learned in one domain to another [94]. This ability to extend the knowledge learned into one context to another reduces the computational cost of the new model, as it requires

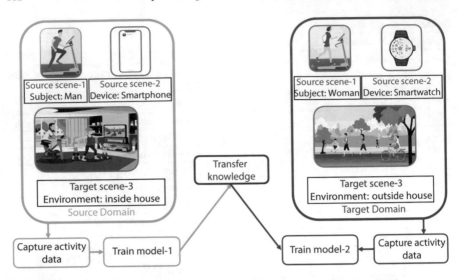

**Fig. 9.10** Basic illustration of transfer learning for different source and target scenarios

less amount of training sample [95]. The idea behind transferring knowledge from one domain to another is to assume the existing relationship between source and target. In Fig. 9.10, we have illustrated the application scenario of transfer learning for sensor-based HAR. In the training phase, the model has been trained to recognize daily human activities using smartphone sensor data for a man inside a house. Now, we can think about a different condition to recognize the activities of a woman, using smartwatch sensor data and in an outdoor environment. Using conventional pattern recognition approaches or deep models, a model trained in the past scenario will not perform well for this changed scenario. In this case, we can utilize the concept of transfer learning to train a new model (Model-2), by transferring the knowledge from Model-1 with less amount of annotated data. This procedure will be computationally efficient.

The different conditions and scenarios of the application of transfer learning were explained in research work [94]—detailing sensor modalities, data labeling procedure, and taxonomy of transferred knowledge. Research work [96] performed their experiment on models with different probability distributions to transfer knowledge among them. They also evaluated the performance on HAR [87], Daily Sports [97], and MHealth [98] datasets. As we mentioned earlier, the variation of the probability distribution of acceleration data for different users can degrade the performance of a model, which has been trained on a different person and tested on another. A cross-person activity recognition model with transfer learning and Reduced Kernel Extreme Learning Machine (RKELM) has been proposed to solve this issue for large datasets [99]. On the other hand, research work [100] has proposed a new framework named Stratified Transfer Learning to transfer knowledge (labeled activity data) from the source domain to the target domain into the same subspace. This method has been evaluated on the OPPORTUNITY [55], PAMAP2 [101], and Daily Sports [97] datasets.

A hierarchical Bayesian transfer learning model has been proposed in [102] to solve the problem of correct labeling of data. They evaluated the model using smartphone dataset [87] as their source and USC-HAD dataset [103] as the target. For a high variety of data with different conditions and scenarios, the research work [104] has proposed a transfer learning model. This model has been validated using statistical hypothesis Kolmogorov–Smirnov and $\chi^2$ goodness of fit test and evaluated on the Walk8, HAR [87], and DaSA [105] datasets. On the other hand, a proposition for independent retraining of machine learning algorithms has been proposed without the requirement of any labeled training data [106]. This model has been evaluated on the OPPORTUNITY dataset [55] and DaSA dataset [105].

The challenging factors for human activity recognition are the type of sensors, environmental and experimental setting, subject, and so on. A trained model using specific settings can be evaluated in other settings using the concept of transfer learning. This is the reason behind the intense interest of researchers to discover the use-case of transfer learning in the large scale cross-domain human activity recognition using different sensor modalities and devices.

## 9.12 Active Learning in Sensor-Based HAR

This is a very new research topic in the field of sensor-based HAR. The primary goal of active learning is to lessen learning complexity and cost. It helps to pick an appropriate number of unlabeled, insightful data samples and to query the annotator for the labels, which helps to the labeling effort to have effective results [17]. The active learning-enabled model approach has been shown in Fig. 9.11.

From this Fig. 9.11, we can see the following steps for an active learning method:

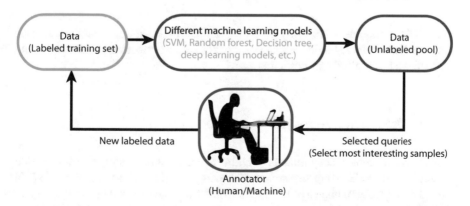

**Fig. 9.11** Basic cycle of active learning approach

- We will be given an unlabeled pool of data.
- We can use existing methods or description logic-based new methods to rank the examples in the order of informativeness.
- Then we need to query the labels of the most informative examples of an unlabeled pool of data.
- The new labeled data are added with the labeled training examples.
- Then the model is retrained using the new training data.

There have been a few pieces of research on active learning to solve the problem of HAR. To find unlabeled data samples with more information and to lessen the effort of annotation, research work [107] has proposed three different techniques (i.e., least confidence method, margin sampling, and entropy-based). On the other hand, k-means clustering algorithm-based active learning approach has been proposed by [17]. This approach used the sampling of uncertainties to identify the most useful samples of unlabeled data. The self-adaptability of activity recognition model was analyzed by [108] during the incorporation of new sensor data. They evaluated their model on bicycle repair [109], car maintenance [110], and OPPORTUNITY [55] datasets. An adaptive framework using active learning was proposed in the research work by [11]. This approach assists in discovering the behavior pattern from high-speed and multidimensional data. This method has been evaluated on the OPPORTUNITY [55], WISDM [11], and smartphone accelerometer [111] datasets. To select unlabeled data for annotations, two different approaches namely expected entropy and query by a committee for active learning have been utilized [112]. This method used random forest-based classifier for activity inference and validated the results on an in-house dataset [112]. Being an emerging field, active learning has numerous scopes in the future in the field of sensor-based HAR.

## 9.13   Challenges in Deep-Model Based HAR and Feasible Solutions

Though deep models solve many problems of shallow learning, these models have some challenges that should be taken care of for better performance. The following subsections highlight the important issues.

### 9.13.1   Real-Time Activity Recognition

This is hard to deploy deep models in real-time applications using smartphone and wearable devices. Existing research works using mobile [42] and smartwatch [60] trained deep models using an offline method on the remote server. The mobile devices utilized the trained model only, which is neither real-time, nor user-friendly in the case of incremental learning. Limited resource and power consumption of mobile and smart devices are other issues in this regard. If we can reduce the communication cost between device and server, and if we can enhance the computational ability of the devices, we can tackle this problem.

### 9.13.2  Performance Limitation in Unsupervised Learning

Previous researches show that deep learning models demonstrate promising performance in the case of labeled data. However, because of the problems of human labeling error, labeling cost, and time consumption, it is required to increase the performance of deep learning for unsupervised activity recognition.

The knowledge extracted from crowd-sourcing can help in this regard to annotate the unlabeled activities [113]. Research can also be done to collect the labels by preserving the security concerns and other limitations. Transfer learning can be another solution in this regard too, by sharing knowledge between activity-related domains [94, 114, 115].

### 9.13.3  Complex Activity Recognition

In real-time to solve real-life problems, researches need to focus on recognizing more high-level and complex activities, which is a challenging task using existing deep learning models. Naturally, complex activities contain more semantics and context information, hence it turns to be more difficult to recognize complex activities. The correlation of signals is ignored by most of the existing methods, which deteriorates the performance. We can use data from hybrid sensors to increase the performance of activity recognition [19]. We can also use context information (e.g., through Wi-Fi, Bluetooth, or GPS) to characterize the situation [116]. This will help to recognize the user state for more specific activities.

### 9.13.4  Robust and Cost-Effective Deep Models

Limitation of computational resources and power consumption issues are some big challenges in the case of sensor-based HAR. Deep models need a good amount of resources that are not available for wearable devices and smartphones. On the other hand, conventional pattern recognition techniques and general neural networks with a lower amount of layers come with the cost of poor performance in spite of being computationally-efficient. Therefore, it is required to design robust and cost-effective deep models.

We can incorporate human-crafted features along with deep features to increase the performance [43]. The task of learning deep features can be made more efficient by providing some pre-knowledge about activity patterns [117]. We can also incorporate shallow models with lower computational cost and deep models with good performance by sharing parameters.

## 9.14   Conclusion

This chapter portrays a summary of the evaluation of deep models over conventional pattern recognition approaches for sensor-based human activity recognition. We have also pointed out the problems of conventional machine learning approaches in terms of poor generalization and shallow learning, that has been solved by employing the automatic feature learning by the deep models. We have also investigated mostly used deep models like CNN, DNN, RNN, DBN, SAE, etc. in this chapter. Finally, we have mentioned the challenges of deep learning in terms of computational cost and discussed transfer learning and active learning for human activity recognition.

## 9.15   Think Further

1. Mention the challenges of traditional pattern recognition approaches to classify activity data.
2. What is deep learning?
3. How deep learnings are evaluated in sensor-based human activity recognition?
4. What are the drawbacks of hand-crafted features?
5. What is shallow learning?
6. Can deep learning increase the performance in sensor-based HAR?
7. Mention some common deep learning models that can be used in sensor-based HAR.
8. What is Deep Neural Network (DNN)?
9. Mention some importance of DNN.
10. What is Convolutional Neural Network (CNN)?
11. Mention some importance of CNN.
12. How to adapt input signal into a CNN network?
13. How can we adjust parameters in the case of a CNN architecture?
14. What is Recurrent Neural Network (RNN)?
15. Mention some importance of RNN.
16. What is vanishing gradient problem?
17. How can Long Short-Term Memory (LSTM) solve the problem of vanishing gradient?
18. How can LSTM extend the memory of the network?
19. What is Stacked Autoencoder (SAE)?
20. Mention some importance of SAE.
21. What is Deep Belief Network (DBN)?

22. Mention some importance of DBN.
23. What is Restricted Boltzmann Machine (RBM)?
24. Mention some importance of RBM.
25. What is hybrid model?
26. Mention some importance of hybrid models?
27. How to choose the best deep model for sensor-based HAR?
28. What is transfer learning?
29. How can transfer learning increase the performance in sensor-based HAR?
30. What is active learning?
31. Mention some importance of active learning in sensor-based HAR.
32. What are the challenges of deep learning methods?

# References

1. Antar, A.D., Ahad, M.A.R., Shahid, O.: Vision-based action understanding for assistive health-care: a short review. In: IEEE CVPR workshop (2019)
2. Ahad, M.A.R.: Vision and sensor based human activity recognition: Challenges ahead (2020)
3. Antar, A.D., Ahmed, M., Ahad, M.A.R.: Challenges in sensor-based human activity recognition and a comparative analysis of benchmark datasets: a review. In: 2019 Joint 8th International Conference on Informatics, Electronics & Vision (ICIEV) and 2019 3rd International Conference on Imaging, Vision & Pattern Recognition (icIVPR), pp. 134–139. IEEE (2019)
4. Ahad, M.A.R.: Motion History Images for Action Recognition and Understanding. Springer Science & Business Media, Berlin (2012)
5. Ahad, M.A.R.: Computer Vision and Action Recognition: A Guide for Image Processing and Computer Vision Community for Action Understanding, vol. 5. Springer Science & Business Media, Berlin (2011)
6. Hossain, T., Islam, M.S., Ahad, M.A.R., Inoue, S.: Human activity recognition using ear-able device. In: Proceedings of the 2019 ACM International Joint Conference on Pervasive and Ubiquitous Computing and Proceedings of the 2019 ACM International Symposium on Wearable Computers, pp. 81–84. ACM (2019)
7. Tazin, T., Hossain, T., Ahad, M.A.R., Inoue, S.: Activity recognition by using lorawan sensor. In: 2018 ACM International Joint Conference on Pervasive and Ubiquitous Computing and the 2018 International Symposium on Wearable Computers (UbiComp/ISWC) (2018)
8. Ahmed, M., Antar, A.D., Ahad, M.A.R.: An approach to classify human activities in real-time from smartphone sensor data. In: 2019 Joint 8th International Conference on Informatics, Electronics Vision (ICIEV) and 2019 3rd International Conference on Imaging, Vision Pattern Recognition (icIVPR), pp. 140–145 (2019)
9. Wang, J., Chen, Y., Hao, S., Peng, X., Lisha, H.: Deep learning for sensor-based activity recognition: a survey. Pattern Recognit. Lett. **119**, 3–11 (2019)
10. Bengio, Y.: Deep learning of representations: looking forward. In: International Conference on Statistical Language and Speech Processing, pp. 1–37. Springer (2013)
11. Yang, J., Nguyen, M.N., San, P.P., Li, X.L., Krishnaswamy, S.: Deep convolutional neural networks on multichannel time series for human activity recognition. In: Twenty-Fourth International Joint Conference on Artificial Intelligence (2015)
12. Saha, S.S., Rahman, S., Rasna, M.J., Mahfuzul Islam, A.K.M., Ahad, M.A.R.: Du-md: an open-source human action dataset for ubiquitous wearable sensors. In: Joint 7th International Conference on Informatics, Electronics & Vision, 2nd International Conference on Imaging, Vision & Pattern Recognition (2018)

13. Hossain, T., Goto, H., Ahad, M.A.R., Inoue, S.: A study on sensor-based activity recognition having missing data. In: 2018 Joint 7th International Conference on Informatics, Electronics & Vision (ICIEV) and 2018 2nd International Conference on Imaging, Vision & Pattern Recognition (icIVPR), pp. 556–561. IEEE (2018)
14. Rasna, M.J., Hossain, T., Inoue, S., Sha, S.S., Rahman, S., Ahad, M.A.R.: Supervised and neural classifiers for locomotion analysis. In: 2018 ACM International Joint Conference on Pervasive and Ubiquitous Computing and the 2018 International Symposium on Wearable Computers (UbiComp/ISWC) (2018)
15. Yang, Q.: Activity recognition: linking low-level sensors to high-level intelligence. In: Twenty-First International Joint Conference on Artificial Intelligence (2009)
16. Zaher, A., Faridee, M., Ramamurthy, S.R., Hossain, H.M., Roy, N.: Happyfeet: recognizing and assessing dance on the floor. In: Proceedings of the 19th International Workshop on Mobile Computing Systems & Applications, pp. 49–54. ACM (2018)
17. Active learning enabled activity recognition: Sajjad Hossain, H.M., Khan, M.A.A.H., Roy, N. Pervasive Mobile Comput. **38**, 312–330 (2017)
18. Hinton, G.E., Osindero, S., Teh, Y.-W.: A fast learning algorithm for deep belief nets. Neural Comput. **18**(7), 1527–1554 (2006)
19. Vepakomma, P., De, D., Das, S.K., Bhansali, S.: A-wristocracy: Deep learning on wrist-worn sensing for recognition of user complex activities. In: 2015 IEEE 12th International Conference on Wearable and Implantable Body Sensor Networks (BSN), pp. 1–6. IEEE (2015)
20. Walse, K.H., Dharaskar, R.V., Thakare, V.M.: Pca based optimal ann classifiers for human activity recognition using mobile sensors data. In: Proceedings of First International Conference on Information and Communication Technology for Intelligent Systems: Volume 1, pp. 429–436. Springer (2016)
21. Hammerla, N.Y., Halloran, S., Plötz, T.: Deep, convolutional, and recurrent models for human activity recognition using wearables (2016). arXiv:1604.08880
22. Bengio, Y. et al.: Learning deep architectures for ai. Found. Trends® Mach. Learn. **2**(1), 1–127 (2009)
23. Deng, L.: A tutorial survey of architectures, algorithms, and applications for deep learning. APSIPA Trans. Signal Inf. Process. **3**, (2014)
24. Fukushima, K.: Neocognitron: a self-organizing neural network model for a mechanism of pattern recognition unaffected by shift in position. Biol. Cybern. **36**(4), 193–202 (1980)
25. Zeng, M., Nguyen, L.T., Yu, B., Mengshoel, O.J., Zhu, J., Wu, P., Zhang, J.: Convolutional neural networks for human activity recognition using mobile sensors. In: 6th International Conference on Mobile Computing, Applications and Services, pp. 197–205. IEEE (2014)
26. Chen, Y., Xue, Y.: A deep learning approach to human activity recognition based on single accelerometer. In: 2015 IEEE International Conference on Systems, Man, and Cybernetics, pp. 1488–1492. IEEE (2015)
27. Pourbabaee, B., Roshtkhari, M.J., Khorasani, K.: Deep convolutional neural networks and learning ecg features for screening paroxysmal atrial fibrillation patients. IEEE Trans. Syst. Man Cybern.: Syst. **99**, 1–10 (2017)
28. Sathyanarayana, A., Joty, S., Fernandez-Luque, L., Ofli, F., Srivastava, J., Elmagarmid, A., Taheri, S. and Arora, T.: Impact of physical activity on sleep: a deep learning based exploration (2016). arXiv:1607.07034
29. Ha, S., Yun, J-M., Choi, S.: Multi-modal convolutional neural networks for activity recognition. In: 2015 IEEE International Conference on Systems, Man, and Cybernetics, pp. 3017–3022. IEEE (2015)
30. Jiang, W., Yin, Z.: Human activity recognition using wearable sensors by deep convolutional neural networks. In: Proceedings of the 23rd ACM international conference on Multimedia, pp. 1307–1310. ACM (2015)
31. Singh, M.S., Pondenkandath, V., Zhou, B., Lukowicz, P. and Liwickit, M.: Transforming sensor data to the image domain for deep learning—an application to footstep detection. In: 2017 International Joint Conference on Neural Networks (IJCNN), pp. 2665–2672. IEEE (2017)

32. Kim, Y., Toomajian, B.: Hand gesture recognition using micro-doppler signatures with convolutional neural network. IEEE Access **4**, 7125–7130 (2016)
33. Zebin, T., Scully, P.J., Ozanyan, K.B.: Human activity recognition with inertial sensors using a deep learning approach. In: 2016 IEEE SENSORS, pp. 1–3. IEEE (2016)
34. Edel, M., Köppe, E.: Binarized-blstm-rnn based human activity recognition. In: 2016 International Conference on Indoor Positioning and Indoor Navigation (IPIN), pp. 1–7. IEEE (2016)
35. Guan, Yu.: Plötz, Thomas: Ensembles of deep lstm learners for activity recognition using wearables. Proceedings of the ACM on Interactive, Mobile, Wearable and Ubiquitous Technologies **1**(2), 11 (2017)
36. Inoue, M., Inoue, S., Nishida, T.: Deep recurrent neural network for mobile human activity recognition with high throughput. Artif. Life Robot. **23**(2), 173–185 (2018)
37. Zeng, M., Gao, H., Yu, T., Mengshoel, O.J., Langseth, H., Lane, I., Liu, X.: Understanding and improving recurrent networks for human activity recognition by continuous attention. In: Proceedings of the 2018 ACM International Symposium on Wearable Computers, pp. 56–63. ACM (2018)
38. Almaslukh, B., AlMuhtadi, J., Artoli, A.: An effective deep autoencoder approach for online smartphone-based human activity recognition. Int. J. Comput. Sci. Netw. Sec. **17**, 160 (2017)
39. Wang, A., Chen, G., Shang, C., Zhang, M., Liu, L.: Human activity recognition in a smart home environment with stacked denoising autoencoders. In: International Conference on Web-Age Information Management, pp. 29–40. Springer (2016)
40. Li, Y., Shi, D., Ding, B., Liu, D.: Unsupervised feature learning for human activity recognition using smartphone sensors. In: Mining intelligence and knowledge exploration, pp. 99–107. Springer (2014)
41. Hammerla, N.Y., Fisher, J., Andras, P., Rochester, L., Walker, R., Plötz, T.: Pd disease state assessment in naturalistic environments using deep learning. In: Twenty-Ninth AAAI Conference on Artificial Intelligence (2015)
42. Lane, N.D., Georgiev, P., Qendro, L.: Deepear: robust smartphone audio sensing in unconstrained acoustic environments using deep learning. In: Proceedings of the 2015 ACM International Joint Conference on Pervasive and Ubiquitous Computing, pp. 283–294. ACM (2015)
43. Plötz, T., Hammerla, N.Y., Olivier, P.L.: Feature learning for activity recognition in ubiquitous computing. In: Twenty-Second International Joint Conference on Artificial Intelligence (2011)
44. Radu, V., Lane, N.D., Bhattacharya, S., Mascolo, C., Marina, M.K., Kawsar, F.: Towards multimodal deep learning for activity recognition on mobile devices. In: Proceedings of the 2016 ACM International Joint Conference on Pervasive and Ubiquitous Computing: Adjunct, pp. 185–188. ACM (2016)
45. Li, X., Zhang, Y., Li, M., Marsic, I., Yang, J., Burd, R.S.: Deep neural network for rfid-based activity recognition. In: Proceedings of the Eighth Wireless of the Students, pp. by the Students, and for the Students Workshop, pp. 24–26. ACM (2016)
46. Fang, H., Hu, C.: Recognizing human activity in smart home using deep learning algorithm. In: Proceedings of the 33rd Chinese Control Conference, pp. 4716–4720. IEEE (2014)
47. Zhang, L., Wu, X., Luo, D.: Real-time activity recognition on smartphones using deep neural networks. In: 2015 IEEE 12th Intl Conf on Ubiquitous Intelligence and Computing and 2015 IEEE 12th Intl Conf on Autonomic and Trusted Computing and 2015 IEEE 15th Intl Conf on Scalable Computing and Communications and Its Associated Workshops (UIC-ATC-ScalCom), pp. 1236–1242. IEEE (2015)
48. Yao, S., Hu, S., Zhao, Y., Zhang, A., Abdelzaher, T.: Deepsense: a unified deep learning framework for time-series mobile sensing data processing. In: Proceedings of the 26th International Conference on World Wide Web, pp. 351–360. International World Wide Web Conferences Steering Committee (2017)
49. Giallanza, T., Siems, T., Smith, E., Gabrielsen, E., Johnson, I., Thornton, M.A., Larson, E.C.: Keyboard snooping from mobile phone arrays with mixed convolutional and recurrent neural networks. Proc. ACM Interact. Mobile Wearab. Ubiquit. Technol. **3**(2), 45 (2019)

50. Ordóñez, F., Roggen, D.: Deep convolutional and lstm recurrent neural networks for multi-modal wearable activity recognition. Sensors **16**(1), 115 (2016)
51. Zheng, Y., Liu, Q., Chen, E., Ge, Y., Zhao, J.L.: Exploiting multi-channels deep convolutional neural networks for multivariate time series classification. Front. Comput. Sci. **10**(1), 96–112 (2016)
52. Liu, C., Zhang, L., Liu, Z., Liu, K., Li, X., Liu, Y.: Lasagna: towards deep hierarchical understanding and searching over mobile sensing data. In: Proceedings of the 22nd Annual International Conference on Mobile Computing and Networking, pp. 334–347. ACM (2016)
53. Wu, Z., Sivadas, S., Tan, Y.K., Bin, M., Goh, R.S.M.: Multi-modal hybrid deep neural network for speech enhancement. *arXiv preprint* arXiv:1606.04750 (2016)
54. Zheng, Y., Liu, Q., Chen, E., Ge, Y., Zhao, J.L.: Time series classification using multi-channels deep convolutional neural networks. In: International Conference on Web-Age Information Management, pp. 298–310. Springer (2014)
55. Roggen, D., Forster, K., Calatroni, A., Holleczek, T., Fang, Y., Troster, G., Ferscha, A., Holzmann, C., Riener, A., Lukowicz, P. et al.: Opportunity: Towards opportunistic activity and context recognition systems. In: 2009 IEEE International Symposium on a World of Wireless, Mobile and Multimedia Networks & Workshops, pp. 1–6. IEEE (2009)
56. Zappi, P., Lombriser, C., Stiefmeier, T., Farella, E., Roggen, D., Benini, L., Tröster, G.: Activity recognition from on-body sensors: accuracy-power trade-off by dynamic sensor selection. In: European Conference on Wireless Sensor Networks, pp. 17–33. Springer (2008)
57. Ronao, C.A., Cho, S.-B.: Human activity recognition with smartphone sensors using deep learning neural networks. Expert Syst. Appl. **59**, 235–244 (2016)
58. Ravi, D., Wong, C., Lo, B., Yang, G.-Z.: A deep learning approach to on-node sensor data analytics for mobile or wearable devices. IEEE J. Biomed. Health Inf. **21**(1), 56–64 (2016)
59. Alsheikh, M.A., Selim, A., Niyato, D., Doyle, L., Lin, S., Tan, H.P.: Deep activity recognition models with triaxial accelerometers. In: Workshops at the Thirtieth AAAI Conference on Artificial Intelligence (2016)
60. Bhattacharya, S., Lane, N.D.: From smart to deep: Robust activity recognition on smart-watches using deep learning. In: 2016 IEEE International Conference on Pervasive Computing and Communication Workshops (PerCom Workshops), pp. 1–6. IEEE (2016)
61. Panwar, M., Dyuthi, S.R., Prakash, K.C., Biswas, D., Acharyya, A., Maharatna, K., Gautam, A., Naik, G.R.: Cnn based approach for activity recognition using a wrist-worn accelerometer. In: 2017 39th Annual International Conference of the IEEE Engineering in Medicine and Biology Society (EMBC), pp. 2438–2441. IEEE (2017)
62. Sani, S., Wiratunga, N., Massie, S.: Learning deep features for knn-based human activity recognition (2017)
63. San, P.P., Kakar, P., Li, X.L., Krishnaswamy, S., Yang, J.B., Nguyen, M.N.: Deep learning for human activity recognition. In: Big Data Analytics for Sensor-Network Collected Intelligence, pp. 186–204. Elsevier (2017)
64. Chen, Y., Zhong, K., Zhang, J., Sun, Q., Zhao, X.: Lstm networks for mobile human activity recognition. In: 2016 International Conference on Artificial Intelligence: Technologies and Applications. Atlantis Press (2016)
65. Cheng, W.Y., Scotland, A., Lipsmeier, F., Kilchenmann, T., Jin, L., Schjodt-Eriksen, J., Wolf, D., Zhang-Schaerer, Y.P., Garcia, I.F., Siebourg-Polster, J. et al.: Human activity recognition from sensor-based large-scale continuous monitoring of parkinson's disease patients. In: Proceedings of the Second IEEE/ACM International Conference on Connected Health: Applications, Systems and Engineering Technologies, pp. 249–250. IEEE Press (2017)
66. Gjoreski, H., Bizjak, J., Gjoreski, M., Gams, M.: Comparing deep and classical machine learning methods for human activity recognition using wrist accelerometer. In: Proceedings of the IJCAI 2016 Workshop on Deep Learning for Artificial Intelligence, New York, NY, USA, vol. 10 (2016)
67. Ha, S., Choi, S.: Convolutional neural networks for human activity recognition using multiple accelerometer and gyroscope sensors. In: 2016 International Joint Conference on Neural Networks (IJCNN), pp. 381–388. IEEE (2016)

68. Hannink, J., Kautz, T., Pasluosta, C.F., Gaßmann, K.-G., Klucken, J., Eskofier, B.M.: Sensor-based gait parameter extraction with deep convolutional neural networks. IEEE J. Biomed. Health Inf. **21**(1), 85–93 (2016)

69. Hayashi, T., Nishida, M., Kitaoka, N., Takeda, K.: Daily activity recognition based on dnn using environmental sound and acceleration signals. In: 2015 23rd European Signal Processing Conference (EUSIPCO), pp. 2306–2310. IEEE (2015)

70. Khan, U.M., Kabir, Z., Hassan, S.A., Ahmed, S.H.: A deep learning framework using passive wifi sensing for respiration monitoring. In: GLOBECOM 2017-2017 IEEE Global Communications Conference, pp. 1–6. IEEE (2017)

71. Kim, Y., Li, Y.: Human activity classification with transmission and reflection coefficients of on-body antennas through deep convolutional neural networks. IEEE Trans. Antennas Propag. **65**(5), 2764–2768 (2017)

72. Lane, N.D., Georgiev, P.: Can deep learning revolutionize mobile sensing? In: Proceedings of the 16th International Workshop on Mobile Computing Systems and Applications, pp. 117–122. ACM (2015)

73. Lee, S-M., Yoon, S.M., Cho, H.: Human activity recognition from accelerometer data using convolutional neural network. In: 2017 IEEE International Conference on Big Data and Smart Computing (BigComp), pp. 131–134. IEEE (2017)

74. Mohammed, S., Tashev, I.: Unsupervised deep representation learning to remove motion artifacts in free-mode body sensor networks. In: 2017 IEEE 14th International Conference on Wearable and Implantable Body Sensor Networks (BSN), pp. 183–188. IEEE (2017)

75. Morales, F.J.O., Roggen, D.: Deep convolutional feature transfer across mobile activity recognition domains, sensor modalities and locations. In: Proceedings of the 2016 ACM International Symposium on Wearable Computers, pp. 92–99. ACM (2016)

76. Murad, A.: Pyun, Jae-Young: Deep recurrent neural networks for human activity recognition. Sensors **17**(11), 2556 (2017)

77. Ronao, C.A., Cho, S-B.: Deep convolutional neural networks for human activity recognition with smartphone sensors. In: International Conference on Neural Information Processing, pp. 46–53. Springer (2015)

78. Wang, J., Zhang, X., Gao, Q., Yue, H., Wang, H.: Device-free wireless localization and activity recognition: A deep learning approach. IEEE Trans. Veh. Technol. **66**(7), 6258–6267 (2016)

79. Abdel-Hamid, O., Mohamed, A.R., Jiang, H., Deng, L., Penn, G., Yu, D.: Convolutional neural networks for speech recognition. IEEE/ACM Trans. Audio Speech Lang. Process. **22**(10), 1533–1545 (2014)

80. Zhang, L., Wu, X., Luo, D.: Human activity recognition with hmm-dnn model. In: 2015 IEEE 14th International Conference on Cognitive Informatics & Cognitive Computing (ICCI* CC), pp. 192–197. IEEE (2015)

81. Zhang, L., Wu, X., Luo, D.: Recognizing human activities from raw accelerometer data using deep neural networks. In: 2015 IEEE 14th International Conference on Machine Learning and Applications (ICMLA), pp. 865–870. IEEE (2015)

82. Zhang, Y., Li, X., Zhang, J., Chen, S., Zhou, M., Farneth, R.A., Marsic, I., Burd, R.S.: Car-a deep learning structure for concurrent activity recognition. In: 2017 16th ACM/IEEE International Conference on Information Processing in Sensor Networks (IPSN), pp. 299–300. IEEE (2017)

83. Zhang, S., Ng, W.W.Y., Zhang, J., Nugent, C.D.: Human activity recognition using radial basis function neural network trained via a minimization of localized generalization error. In: International Conference on Ubiquitous Computing and Ambient Intelligence, pp. 498–507. Springer (2017)

84. Hassan, M.M., Uddin, M.Z., Mohamed, A., Almogren, A.: A robust human activity recognition system using smartphone sensors and deep learning. Future Gener. Comput. Syst. **81**, 307–313 (2018)

85. Reyes-Ortiz, J.-L., Oneto, L., Samà, A., Parra, X., Anguita, D.: Transition-aware human activity recognition using smartphones. Neurocomputing **171**, 754–767 (2016)

86. Avilés-Cruz, C., Ferreyra-Ramírez, A., Zúñiga-López, A., Villegas-Cortéz, J.: Coarse-fine convolutional deep-learning strategy for human activity recognition. Sensors **19**(7), 1556 (2019)
87. Anguita, D., Ghio, A., Oneto, L., Parra, X., Reyes-Ortiz, J.L.: A public domain dataset for human activity recognition using smartphones. In: Esann (2013)
88. Antar, A.D., Ahmed, M., Ishrak, M.S., Ahad, M.A.R.: A comparative approach to classification of locomotion and transportation modes using smartphone sensor data. In: Proceedings of the 2018 ACM International Joint Conference and 2018 International Symposium on Pervasive and Ubiquitous Computing and Wearable Computers, pp. 1497–1502 (2018)
89. Gjoreski, H., Ciliberto, M., Wang, L., Morales, F.J.O., Mekki, S., Valentin, S., Roggen, D.: The university of sussex-huawei locomotion and transportation dataset for multimodal analytics with mobile devices. IEEE Access **6**, 42592–42604 (2018)
90. Saha, S.S., Rahman, S., Haque, Z.R.R., Hossain, T., Inoue, S., Ahad, M.A.R.: Position independent activity recognition using shallow neural architecture and empirical modeling. In: Adjunct Proceedings of the 2019 ACM International Joint Conference on Pervasive and Ubiquitous Computing and Proceedings of the 2019 ACM International Symposium on Wearable Computers, pp. 808–813 (2019)
91. Antar, A.D., Ahmed, M., Hossain, T., Muramatsu, D., Makihara, Y., Inoue, S., Yagi, Y., Ahad, M.A.R., Ngo, T.T.: Wearable sensor-based gait analysis for age and gender estimation (2020)
92. Ngo, T.T., Ahad, M.A.R., Antar, A.D., Ahmed, M., Muramatsu, D., Makihara, Y., Yagi, Y., Inoue, S., Hossain, T. and Hattori, Y.: Ou-isir wearable sensor-based gait challenge: age and gender. In: Proceedings of the 12th IAPR International Conference on Biometrics, ICB (2019)
93. Ngo, T.T., Makihara, Y., Nagahara, H., Mukaigawa, Y., Yagi, Y.: The largest inertial sensor-based gait database and performance evaluation of gait-based personal authentication. Pattern Recognit. **47**(1), 228–237 (2014)
94. Cook, D., Feuz, K.D., Krishnan, N.C.: Transfer learning for activity recognition: a survey. Knowl. Inf. Syst. **36**(3), 537–556 (2013)
95. Byrnes, J.P.: Cognitive Development and Learning in Instructional Contexts. Allyn and Bacon Boston (1996)
96. Khan, M.A.A.H., Roy, N.: Transact: transfer learning enabled activity recognition. In: 2017 IEEE International Conference on Pervasive Computing and Communications Workshops (PerCom Workshops), pp. 545–550. IEEE (2017)
97. Barshan, B., Yüksek, M.C.: Recognizing daily and sports activities in two open source machine learning environments using body-worn sensor units. Comput. J. **57**(11), 1649–1667 (2014)
98. Banos, O., Garcia, R., Holgado-Terriza, J.A., Damas, M., Pomares, H., Rojas, I., Saez, A., Villalonga, C.: Mhealthdroid: a novel framework for agile development of mobile health applications. In: International Workshop on Ambient Assisted Living, pp. 91–98. Springer (2014)
99. Deng, W.-Y., Zheng, Q.-H., Wang, Z.-M.: Cross-person activity recognition using reduced kernel extreme learning machine. Neural Netw. **53**, 1–7 (2014)
100. Wang, J., Chen, Y., Hu, L., Peng, X., Philip, S.Y.: Stratified transfer learning for cross-domain activity recognition. In: 2018 IEEE International Conference on Pervasive Computing and Communications (PerCom), pp. 1–10. IEEE (2018)
101. Reiss, A., Stricker, D.: Introducing a new benchmarked dataset for activity monitoring. In: 2012 16th International Symposium on Wearable Computers, pp. 108–109. IEEE (2012)
102. Diethe, T., Twomey, N., Flach, P.A.: Active transfer learning for activity recognition. In: ESANN (2016)
103. Zhang, M., Sawchuk, A.A.: Usc-had: a daily activity dataset for ubiquitous activity recognition using wearable sensors. In: Proceedings of the 2012 ACM Conference on Ubiquitous Computing, pp. 1036–1043. ACM (2012)
104. Ying, J.J-C., Lin, B-H., Tseng, V.S., Hsieh, S-Y.: Transfer learning on high variety domains for activity recognition. In: Proceedings of the ASE BigData & Social Informatics 2015, p. 37. ACM (2015)

105. Altun, K., Barshan, B., Tunçel, O.: Comparative study on classifying human activities with miniature inertial and magnetic sensors. Pattern Recognit. **43**(10), 3605–3620 (2010)
106. Rokni, S.A., Ghasemzadeh, H.: Synchronous dynamic view learning: a framework for autonomous training of activity recognition models using wearable sensors. In: Proceedings of the 16th ACM/IEEE International Conference on Information Processing in Sensor Networks, pp. 79–90. ACM (2017)
107. Alemdar, H., van Kasteren, T.L.M., Ersoy, C.: Active learning with uncertainty sampling for large scale activity recognition in smart homes. J. Ambient Intell. Smart Environ. **9**(2), 209–223 (2017)
108. Bannach, D., Jänicke, M., Rey, V.F., Tomforde, S., Sick, B., Lukowicz, B.: Self-adaptation of activity recognition systems to new sensors (2017). arXiv:1701.08528
109. Ogris, G., Stiefmeier, T., Junker, H., Lukowicz, P., Troster, G.: Using ultrasonic hand tracking to augment motion analysis based recognition of manipulative gestures. In: Ninth IEEE International Symposium on Wearable Computers (ISWC'05), pp. 152–159. IEEE (2005)
110. Stiefmeier, T., Roggen, D., Ogris, G., Lukowicz, P., Tröster, G.: Wearable activity tracking in car manufacturing. IEEE Pervasive Comput. **2**, 42–50 (2008)
111. Do, T.M., Loke, S.W., Liu, F.: Healthylife: an activity recognition system with smartphone using logic-based stream reasoning. In: International Conference on Mobile and Ubiquitous Systems: Computing, Networking, and Services, pp. 188–199. Springer (2012)
112. Bagaveyev, S., Cook, D.J.: Designing and evaluating active learning methods for activity recognition. In: Proceedings of the 2014 ACM International Joint Conference on Pervasive and Ubiquitous Computing: Adjunct Publication, pp. 469–478. ACM (2014)
113. Prelec, D., Seung, H.S., McCoy, J.: A solution to the single-question crowd wisdom problem. Nature **541**(7638), 532 (2017)
114. Pan, S.J., Yang, Q.: A survey on transfer learning. IEEE Trans. Knowl. Data Eng. **22**(10), 1345–1359 (2009)
115. Wang, J., Chen, Y., Hao, S., Feng, W., Shen, Z.: Balanced distribution adaptation for transfer learning. In: 2017 IEEE International Conference on Data Mining (ICDM), pp. 1129–1134. IEEE (2017)
116. Abowd, G.D., Dey, A.K., Brown, P.J., Davies, N., Smith, M., Steggles, P.: Towards a better understanding of context and context-awareness. In: International Symposium on Handheld and Ubiquitous Computing, pp. 304–307. Springer (1999)
117. Stewart, R., Ermon, S.: Label-free supervision of neural networks with physics and domain knowledge. In: Thirty-First AAAI Conference on Artificial Intelligence (2017)

# Chapter 10
# Sensor-Based Human Activity Recognition: Challenges Ahead

**Abstract** Human Activity Recognition (HAR) has explored a lot recently in the academia and industries for numerous applications. There are lots of progress in the domain of vision-based action or activity recognition due to the advent of deep learning models and due to the availability of very large datasets in the last several years. However, there are still a number of genuine challenges in vision-based HAR. On the other hand, sensor-based HAR has more constraints to decipher, and is still far from maturity due to various challenges. These core challenging issues are addressed in this chapter. The challenges regarding data collection issues have been discussed in detail. Prospective research works and challenges in the field of sensor-based activity recognition have been discussed in terms of new researchers perspective, possible developments in industries for experts, and smart IoT solutions in the medical sector.

## 10.1  Major Challenges in Sensor-Based HAR

Sensor-based Human Activity Recognition (HAR) has explored a lot recently in the academia and industries for numerous applications. The field is still not matured due to various challenges [1–5]. These core challenging issues are discussed in this chapter. In this chapter, we are not considering the challenges of activity recognition based on video sensors, though some of these issues overlap with the constraints of sensor-based human activity recognition and analysis [2, 6, 7].

The current and possible future challenges regarding data collection for sensor-based human activity recognition (HAR) have been summarized in Fig. 10.1. From this Fig. 10.1, the challenges are summarized below:

- Diversity of age, gender, and number of subjects
- Postural transitions
- Number of sensors and types of sensors
- Different body locations of wearable sensors or smartphone
- Missing values or labeling error
- Similar postures

© Springer Nature Switzerland AG 2021
M. A. R. Ahad et al., *IoT Sensor-Based Activity Recognition*, Intelligent Systems Reference Library 173, https://doi.org/10.1007/978-3-030-51379-5_10

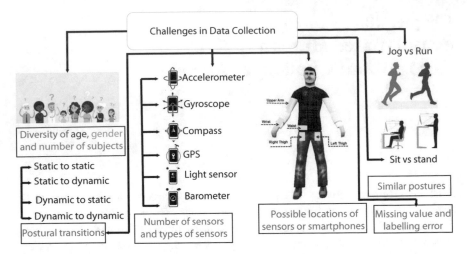

**Fig. 10.1**  Some core challenges in the case of data collection for sensor-based HAR research

- Datasets having complex activities
- Datasets having multiple persons and their mutual interactions
- Datasets having fall down-related activities
- Datasets having realistic scenes by the genuine age groups, instead of young adults for all activity classes.

### 10.1.1   Diversity of Age, Gender, and Number of Subjects

There are several challenges associated with data collection. First of all, there should be diversity in terms of age, gender, and the number of subjects while creating a dataset. A dataset is regarded as a benchmark dataset, if the dataset represents the real-world scenario. It is possible to develop if the dataset contains data from people of different ages and genders with a high amount of participation. We can mention the OU-ISIR Gait Database, Inertial Sensor Dataset [8–10] as an example of a diverse dataset. This dataset contains 745 subjects, where 388 are males and 357 are females (and the age range is from 2 to 78 years, though the number of elderly people were less).

### 10.1.2   Postural Transitions

We can see that tracking activities with postural transitions can also be difficult and it is very hard to track the change of postures using sensors while switching from one activity to another [11]. For example, there is a static to dynamic postural transition when a person starts walking from still positions, whereas, in between starting running from walking activity, there is a dynamic to dynamic postural transition. We can increase the recognition rate if we can track the change of postures while switching from one activity to another, which is a difficult task. The switching occurs very fast and this is the reason for the difficulty to track the transition in an accurate manner.

### 10.1.3 Number of Sensors and Types of Sensors

There are various types of sensors that are available for collecting data. Nowadays, a smartphone or a smartwatch also contains different kinds of sensors like accelerometer, gyroscope, GPS, pressure sensor, light sensor, etc. These smart devices, wearable devices, or environmental sensors can be used for collecting human activity data. The challenges are to fix the number of sensors and type of sensors to get better recognition results. We need to carefully choose the number of sensors as in real-life implementations a recognition device that has been implemented with a limited collection of sensors allows the task simpler and easy. The number of sensors plays a significant role in the case of wearable sensors due to user-comfort concerns, as well as data density or even data redundancy problems. Carrying multiple sensors is difficult for users. However, one device may contain multiple sensors (e.g., TI SensorTag, earable devices [5], and so on). If we can exploit some of these sensors smartly and application-wise then the results will be better. There is a trade-off between the number of sensors and the efficiency to be treated carefully [12].

### 10.1.4 Different Body Locations of Wearable Sensors or Smartphone

Locations of smartphones or wearable sensors also play an important role in HAR. Often we can see that a model trained with sensor data keeping the smartphone in the waist position fails to give good performance if we use the same model to predict activities for sensor data by keeping the smartphone in hand or bag [13]. Location-, position-, and orientation-invariant approaches are required for robust applications [9, 10, 14, 15]. Another problem is that some users may not keep their phones with them while they are staying at home, making tracking their activities impossible. In this case, a wearable sensor can be a good option for many users to wear it all day long while performing activities, although it comes with the problem of discomfort.

## 10.1.5   Missing Value

Another challenge is to deal with missing values from the sensors, or in the dataset. Missing values can occur due to different factors, e.g.,

- Limitation of computational resources in mobile devices,
- Optimization problem of the wireless sensor network,
- Malfunction of sensors,
- Data packet loss or collision,
- Distance between the sensors and access points,
- Synchronization problem of sensors,
- Weak Wi-Fi signal or poor network coverage,
- Environmental noises, and so on.

Missing data can deteriorate the performance of human activity recognition unless the missing data can be imputed/augmented or handled differently [3, 16–22].

## 10.1.6   Labeling Error

Apart from the missing data issue, labeling error also causes bad performance while the dataset is labeled manually. We can utilize machine learning techniques to deal with these challenges. Moreover, we can think of implementing unsupervised learning using unlabeled data, which can be an alternative of labeled data with an error (having considerable trade-off). We can also think of automatic annotation avoiding the general practice of manual annotation by a human [23].

## 10.1.7   Similar Postures

Similar postures like sitting and stand, or jog and run make it difficult to recognize the activity. Suppose, there are some persons who sit on a sofa in such a way, which is very similar to sleeping posture. This case can also happen for dynamic activities like jog and run, which seem similar in terms of posture. There are some research going on to identify the action distinguishing ambiguous postures [24].

## 10.2   Challenges Ahead: Headway and Diversity

Future research works and challenges in the field of activity recognition can be discussed in terms of three perspectives such as (Fig. 10.2),

- Challenges for new researchers,

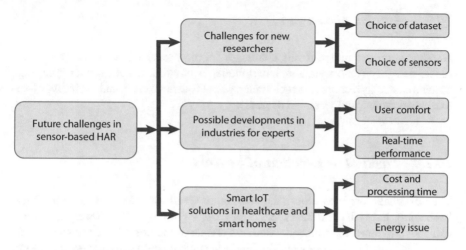

**Fig. 10.2** Challenges ahead for the researchers in terms of three basic perspectives

- Possible developments in industries for experts, and
- Smart IoT solutions in healthcare and smart homes.

## 10.2.1  Selection of Appropriate Datasets

For a new researcher, it is quite very much difficult to choose the perfect dataset that is more related to the research field. In this book, we have presented more than 150 datasets regarding numerous fields of activity recognition along with activity labels, sensors, and device information for researchers. A summarized version on datasets are available in [25]. Researchers who are willing to create sensor-based datasets often find difficulties in terms of sensor choice, the number of subjects, varieties in the dataset in terms of gender and age of the subjects, ethical factors regarding users while data collection and so on.

## 10.2.2  Lack of Ground Truths

Some databases for the researchers do not include ground-truth and precise knowledge. Hence, a researcher may not conceive and analyze beyond the conventional methods in analyzing different activities.

### 10.2.3 Choice of Sensors

Another common and difficult challenge for new researchers is to choose the sensor category (e.g., wearable, smartphone, or environmental), sensor type (e.g., accelerometer, gyroscope, magnetometer, pressure sensor, etc.), and device type (various smartphones and wearable devices).

### 10.2.4 Position or Location of Sensors

The next challenge is to choose the position of wearable sensors (e.g., upper back, back waist, left wrist, left lower leg, right wrist, right thigh, right lower leg). While using smartphones for data collection, the choice of the position of the smartphone (e.g., shirt's pocket, trouser's pocket, inside a handbag, etc.) also matters in terms of accuracy and precision.

### 10.2.5 User Comfort Issue

When human activity research is performed to develop products in the industry that may assist and help people in their daily life, we need to focus on many parameters and precautions. User comfort and familiarity of the chosen device also play important roles in this case [26]. Otherwise, even if a particular sensor position brings higher accuracies, we cannot choose that position ignoring user comfort.

### 10.2.6 Processing Sensor Data and Feature Vector

The early challenges for the new researchers after data collection include visualization of data, choice of correct filtering method for preprocessing, choice of window length, type and overlap percentage for segmentation. Choice of more important features using a robust method also creates additional challenges for the researchers to minimize the processing time in real-time keeping good accuracy percentage.

### 10.2.7 Selection of Classifiers

The choice of classifiers and choosing between general classifiers and deep learning techniques need to be handled more carefully for building a good robust model without overfitting. Model size, processing time, data type, accuracy, precision, application areas—all these play crucial functions in this regard.

A group of classifier-based approach may be used for the proper identification of certain related behaviors such as sitting and standing or walking, going upstairs and going downstairs, or leaning backward and sitting backward in a rolling chair. Much of the current works struggled to specifically differentiate related behaviors with one classifier. It is necessary to explore the ensemble of classifiers with voting for the classification of complex activities [14, 27–34]. Future planning will also involve dealing with manually labeled training data which is not correctly classified as walking exercise but mistakenly labeled as jogging exercise due to human error.

### 10.2.8 Analysis of Resource Consumption and Real-Time Assessment

The study of resource use, like memory, CPU, number of sensors, and, most notably, battery use, should be performed in potential research work [35–42]. The most common trade-off between the recognition accuracy and precision and resource usage should be explored more deeply to get the best possible outcome by the experts. Some activities are required to understand in real-time. Understanding or recognizing or detecting fall down is very important activity for the elderly people and hospital patients [43]. Falling down becomes one of the serious concerns in hospitals in Japan, New Zealand and similar other countries where the elderly population are higher in percentage. To ensure real-time processing, computing capacity should be better. However, we can not ensure higher computation in wearable devices. This area is important and up to the industry to handle the related challenges.

### 10.2.9 Variability and Diversity in Datasets

The datasets should have more variabilities so that the challenges become more realistic and more application-centric. Some issues are highlighted in the following points:

- One of the main problems of most of the currently available activity recognition databases is less amount of users without varieties (i.e., age-range, gender, environment, etc.).
- Besides this, most of the datasets only focus on general and simple human activities like walk, run, sit, etc. More advanced datasets with more complex activity labels (e.g., eating with hand, eating with a spoon, brushing teeth, walking upstairs, etc.) with more variation of users should be created in future research works.
- Besides, most of the datasets are created inside the lab under controlled environments by trained users, which does not perform well in real-life environment. Researchers should intend to create more realistic datasets considering these problems so that the training dataset can represent real-life environmental situations.

## 10.2.10   Smart IoT Solutions in Healthcare and Smart Homes

Human activity recognition research has vast application areas in healthcare and building smart homes with the advancement of smart sensing devices and miniaturized smart sensors with the power of the Internet of Things (IoT) [44]. Researchers should explore more areas for making activity related datasets that can improve the medical sector and smart home system. Some of these can be,

- Nursing activities related dataset,
- Patient activities and monitoring dataset (e.g., during the hospital, staying home and monitoring, monitoring after a surgery and the improvement assessment),
- Abnormal activities and fall detection related datasets,
- Physical exercise-related dataset,
- Elderly activity monitoring dataset,
- Pregnant woman activities monitoring,
- Children's behavior assessment-related dataset,
- Autism study-related dataset [45], etc.

In this field of research, the two important facts are sensor cost and energy issue. Most of the patients do not find it comfortable to wear heavy weight and large-sized sensing devices with lots of wires. This is important to enable wireless sensing technology with light weight sensors and miniaturized devices. The utilization of nanotechnology in this regard makes the sensors costly. The use of small-sized batteries with limited energy backup also poses a challenge to continue the data collection process. Nowadays, researchers focus on designing low dimensional features and models with lower computational costs that can be implemented on a small device with limited energy resources [46, 47]. In spite of challenges, we can not deny the importance of sensor-based activity recognition research in the healthcare domain [1, 48].

## 10.2.11   Multi-modal Options

In future, we need to incorporate various other modalities—especially from the vision- and image-based domain [2], such as,

- Gaze and attention analysis,
- Voice or speech signal processing,
- Social signal analysis,
- Visual data for human behavior analysis,
- Facial attributes to understand any objective or purpose of an action, not just the movement's information and decision,
- Collective human behavior and data analysis in the multi-person cases or interactions.

## 10.2.12 Incorporation of Voice or Audio Data

Regarding voice or speech, there is a new device called *eSense earable* by Nokia-Bell Lab, UK. It has a multi-modal stereo system along with inertial measurement unit (IMU). This device can incorporate motion, audio, and proximity data. Activity analysis are explored with this device recently by [5]. It is found that visual frames and audio can be effective to enhance the recognition results in the case of large-scale video classification [49]. However, in the domain of sensor-based activity, the explorations as well as the success are insignificant. We need to explore the audio along with sensor data. In our life, we do many activities while talking with others or using mobile phones. Audio can give the clue on human personal emotional status like happy or sad or anger or others. A movement along with any emotional status can bring better recognition results.

## 10.2.13 Large-Scale and Multi-label Case

Large-scale and multi-label activity classification is a major challenge for the future. For this kind of case, fusion of different modalities is a difficult task. We need to explore approaches for these issues. Vision-based activity and behavior understanding on large-scale cases are advanced in the last few years, compared with the sensor-based domain. Therefore, it is required to explore the vision-based domain and learn from the experiences and approaches—so that we can accommodate some of these in sensor-based activity analysis.

## 10.2.14 Person-Object Activities

Objection recognition is not done in the sensor domain. Video or image-based object recognition has progressed a lot. Now, can we explore sensors in the domain of person and object interactions? For example, reading book, opening a door, playing guitar, cooking something, preparing bed, cleaning floors, moving an object, etc. are very much done in daily life. Open a door of a home, or opening the refrigerator, or opening the door of a car—these vary object-wise and hence, *context/scene* analysis becomes necessary for this kind of activities.

## 10.2.15 Hand-Object Interaction

Hand-object interaction using sensor information can be another important research area. A hand can grip different objects and hence, the differences should be understood automatically. Context-awareness is essential for activity understanding.

Hand-object interaction can be very useful in the study of partially paralyzed patients and during their rehabilitation progress periods—to understand how much the developments are made. There can be other applications on this arena. However, how to locate the sensors and how to accommodate the collected data can be difficult and challenging issue. Application-based realistic options can be considered in future in different health-care and rehabilitation applications.

### 10.2.16   Person-Person Interaction

Most of the datasets we presented in this book are basically *singular* person actions/activities (e.g., walking, running, hiking, etc.). However, person-person actions or *interactions* are more important to understand. Two persons can handshake, hug, fight, push, kiss, discuss, etc. More than two persons can make more complex activities. These are really challenging tasks using sensor data only—because we need to explore the data from both or multiple persons in the vicinity to understand the actions or interactions. Nevertheless, it is essential to explore these interactions to understand many real-life applications.

### 10.2.17   Errors in Data Collection and Missing Values

Data missing is a realistic issue and challenging topic. Due to weak WiFi connection or distance or low battery level or other technical issues in a wireless sensor network or body area network, data can be missed or lost in different statistical patterns [16]. Hence, research work should be intended in future to correct identification of a particular activity in real-time that has missing values in training dataset due to the failure of the central server to receive sensor data or that includes noisy data due to sensor intervention while performing the activity.

In addition, researchers can focus to design a feedback-based network that will have the capability of continuous improvement of its learning algorithm through the assessment of users. Whenever a wrong identification of any activity is assured by a user, this network will take this information and upgrade its learning so that it can correctly identify that activity more precisely in the future.

### 10.2.18   Challenges Imposed by Deep Models

Nowadays, many research works focus on deep models avoiding shallow hand-crafted features to make the model generalized and to transfer the knowledge among multiple domains. Several factors have also been found by the researchers that must be taken care of in the future to maintain better performance while using deep learning-based approaches. Most of the previous approaches that utilize deep-learning techniques are offline-based techniques using pre-trained models. However, this approach may fail to perform well in real-time for smartphones and smart wearable devices with limited energy resource. Besides, deep learning-based models can pose a challenge to obtain better performance for unlabeled data in the case of unsupervised learning. Identification of complex activities consisted of several sub-activities or basic actions is difficult using current deep models because of the presence of semantics and context information. Moreover, research should be focused to design deep models with lower computational cost keeping a balanced performance so that we can implement those models in miniaturized sensing devices with the lower computational ability and limited energy resources.

## 10.3 Concluding Remarks

With the rapid development and technical progress in the field of IoT sensors, behavior detection in a variety of evolving computing fields has been the latest frontier of context-aware customized applications. But the reality is there are not many systematic studies in sensor-based behavior detection, which is the reason behind this field becoming a new area of study. Scientists do not find standard datasets most of the time, which may render their research incredibly difficult, even in practical environments.

This book (in Chap. dbSen6) surveys the state-of-the-art human activity recognition where we have compiled more than 150 sensor-based benchmark datasets on daily activities, ambulation activities, medical activities, fitness activities, wearable sensor-based, and smartphone sensor-based activities. Besides benchmark datasets on various types of fall detection techniques have been presented with relevant information. Specific details were presented about the characteristics, levels of operation, categories of sensors, and equipment used by these datasets. We also generated a list of all sorts of sensing instruments and implementation software that can be used to create a new dataset. Various noise filtering methods, filter preference, segmentation methods, and considerations to be included in determining the duration of the window have been defined in depth. In comparison, each of these databases provided a description and detailed study of historical behavior identification techniques.

Finally, for future research, numerous ideas are proposed to elongate this field to more practical and pervasive scenarios. This chapter highlighted the core constraints of the existing works. It elaborated a number of future challenges and several notes on possible solutions and ideas for researchers. We conclude this book with the high-hope that this book will be instrumental for the IoT sensor-based research activities that will create brighter future in different applications, especially in healthcare domain.

## 10.4   Think Further

1. Mention some challenges that are still existing in the field of sensor-based human activity recognition.
2. List up the major approaches that you feel the best-suited for your target application or research area.
3. What are the major constraints of your enlisted methods to solve the problems of HAR field?
4. Can you utilize existing numerous concepts to update them or bring a minor change to solve each of your enlisted constraints?
5. How can you get an idea from the existing research works to solve the problems?
6. What are the future challenges that are yet to be solved in the area of human activity recognition?
7. What are the future challenges of data collection protocol?
8. Why diversity of age and gender is necessary for a good dataset?
9. Why postural transition-based activities are difficult to recognize?
10. How can you limit the number of sensors by keeping moderate performance?
11. How to build a position-independent model so that in real-time the user does not require to keep the smartphone in a constant body position?
12. Why similar postures are difficult to classify?
13. How to deal with the problem of missing values while collecting data?
14. What is the solution to missing value problem?
15. How to deal with the problem of human labeling error?
16. What alternate approach can be taken to automatically annotate data?
17. Can unsupervised learning help us to utilize unlabeled data to solve human labeling error problem?
18. Compare the vision-based HAR and the sensor-based HAR based on other vision-based literature and books.
19. What are the intriguing points of sensor-based HAR that can be implemented in the vision-based activity recognition?
20. What are the intriguing features of vision-based HAR that can be considered in the sensor-based HAR?
21. How can we combine some of the features of vision-based and sensor-based activity recognition?
22. Enlist all possible applications where we can explore various IoT sensors and cameras (whether RGB or depth).
23. Draw the workable architecture and flow diagrams for each of the above-questioned applications.
24. If you explore multiple sensors or cameras for one person, what are the challenges?
25. How can you collect data in a synchronized manner so that multi-sensors or cameras can be handled properly?
26. How can you create datasets related to elderly people without engaging or harming them while creating the dataset? What are the challenges?

27. How can you create a dataset related to fall detection?
28. What are the different kinds of falls that can be considered? Note that falling down by a patient or an elderly person may not have the same nature with others.
29. Find scopes of IoT sensor-based activity and behavior understanding in the field of autism.
30. Find applications and scopes of sensor-based HAR related to criminal investigation and forensic applications.

# References

1. Antar, A.D., Ahad, M.A.R., Shahid, O.: Vision-based action understanding for assistive healthcare: a short review. In: IEEE CVPR Workshop (2019)
2. Ahad, M.A.R.: Vision and sensor based human activity recognition: challenges ahead (2020)
3. Hossain, T., Goto, H., Ahad, M.A.R., Inoue, S.: A study on sensor-based activity recognition having missing data. In: 2018 Joint 7th International Conference on Informatics, Electronics & Vision (ICIEV) and 2018 2nd International Conference on Imaging, Vision & Pattern Recognition (icIVPR), pp. 556–561. IEEE (2018)
4. Tazin, T., Hossain, T., Ahad, M.A.R., Inoue, S.: Activity recognition by using lorawan sensor. In: 2018 ACM International Joint Conference on Pervasive and Ubiquitous Computing and the 2018 International Symposium on Wearable Computers (UbiComp/ISWC), 2018
5. Hossain, T., Islam, M.S., Ahad, M.A.R., Inoue, S.: Human activity recognition using earable device. In: Proceedings of the 2019 ACM International Joint Conference on Pervasive and Ubiquitous Computing and Proceedings of the 2019 ACM International Symposium on Wearable Computers, pp. 81–84. ACM, 2019
6. Ahad, M.A.R.: Motion History Images for Action Recognition and Understanding. Springer Science & Business Media, Berlin (2012)
7. Ahad, M.A.R.: Computer Vision and Action Recognition: A Guide for Image Processing and Computer Vision Community for Action Understanding, vol. 5. Springer Science & Business Media, Berlin (2011)
8. Ngo, T.T., Makihara, Y., Nagahara, H., Mukaigawa, Y., Yagi, Y.: The largest inertial sensor-based gait database and performance evaluation of gait-based personal authentication. Pattern Recognit. 47(1), 228–237 (2014)
9. Ngo, T.T., Ahad, M.A.R., Antar, A.D., Ahmed, M., Muramatsu, D., Makihara, Y., Yagi, Y., Inoue, S., Hossain, T., Hattori, Y.: Ou-isir wearable sensor-based gait challenge: age and gender. In: Proceedings of the 12th IAPR International Conference on Biometrics, ICB (2019)
10. Antar, A.D., Ahmed, M., Hossain, T., Muramatsu, D., Makihara, Y., Inoue, S., Yagi, Y., Ahad, M.A.R., Ngo, T.T.: Wearable sensor-based gait analysis for age and gender estimation (2020)
11. Reyes-Ortiz, J.-L., Oneto, L., Ghio, A., Samá, A., Anguita, D., Parra, X.: Human activity recognition on smartphones with awareness of basic activities and postural transitions. In: International Conference on Artificial Neural Networks, pp. 177–184. Springer (2014)
12. Voicu, R.-A., Dobre, C., Bajenaru, L., Ciobanu, R.-I.: Human physical activity recognition using smartphone sensors. Sensors 19(3), 458 (2019)
13. Xu, W., Zhang, M., Sawchuk, A.A., Sarrafzadeh, M.: Co-recognition of human activity and sensor location via compressed sensing in wearable body sensor networks. In: 2012 Ninth International Conference on Wearable and Implantable Body Sensor Networks, pp. 124–129. IEEE (2012)
14. Ahmed, M., Antar, A.D., Hossain, T., Inoue, S., Ahad, M.A.R.: Poiden: position and orientation independent deep ensemble network for the classification of locomotion and transportation modes. pp. 674–679 (2019)

15. Saha, S.S., Rahman, S., Haque, Z.R.R., Hossain, T., Inoue, S., Ahad, M.A.R.: Position independent activity recognition using shallow neural architecture and empirical modeling. In: Adjunct Proceedings of the 2019 ACM International Joint Conference on Pervasive and Ubiquitous Computing and Proceedings of the 2019 ACM International Symposium on Wearable Computers, pp. 808–813 (2019)

16. Ahad, M.A.R., Hossain, T., Tazin, T., Inoue, S.: Study of lorawan technology for activity recognition. In: 2018 ACM International Joint Conference on Pervasive and Ubiquitous Computing and the 2018 International Symposium on Wearable Computers (UbiComp/ISWC) (2018)

17. Chavarriaga, R., Sagha, H., Calatroni, A., Digumarti, S.T., Tröster, G., Millán, J.D.R., Roggen, D.: The opportunity challenge: a benchmark database for on-body sensor-based activity recognition. Pattern Recognit. Lett. **34**(15), 2033–2042 (2013)

18. Akhavian, R., Behzadan, A.: Wearable sensor-based activity recognition for data-driven simulation of construction workers' activities. In: 2015 Winter Simulation Conference (WSC), pp. 3333–3344. IEEE (2015)

19. Yin, J., Yang, Q., Pan, J.J.: Sensor-based abnormal human-activity detection. IEEE Trans. Knowl. Data Eng. **20**(8), 1082–1090 (2008)

20. Wang, L., Gu, T., Tao, X., Lu, J.: Sensor-based human activity recognition in a multi-user scenario. In: European Conference on Ambient Intelligence, pp. 78–87. Springer (2009)

21. Pham, C., Diep, N.N., Phuong, T.M.: A wearable sensor based approach to real-time fall detection and fine-grained activity recognition. J. Mobil. Multimed. **9**(1&2), 15–26 (2013)

22. Tao, G., Wang, L., Zhanqing, W., Tao, X., Jian, L.: A pattern mining approach to sensor-based human activity recognition. IEEE Trans. Knowl. Data Eng. **23**(9), 1359–1372 (2010)

23. Cruciani, F., Cleland, I., Nugent, C., McCullagh, P., Synnes, K., Hallberg, J.: Automatic annotation for human activity recognition in free living using a smartphone. Sensors **18**(7), 2203 (2018)

24. Liu, Z., Yin, J., Li, J., Wei, J., Feng, Z.: A new action recognition method by distinguishing ambiguous postures. Int. J. Adv. Robot. Syst. **15**(1), 1729881417749482 (2018)

25. Antar, A.D., Ahmed, M., Ahad, M.A.R.: Challenges in sensor-based human activity recognition and a comparative analysis of benchmark datasets: A review. In: 2019 Joint 8th International Conference on Informatics, Electronics & Vision (ICIEV) and 2019 3rd International Conference on Imaging, Vision & Pattern Recognition (icIVPR), pp. 134–139. IEEE (2019)

26. Knight, J.F., Baber, C.: A tool to assess the comfort of wearable computers. Hum. Factors **47**(1), 77–91 (2005)

27. Lara, O.D., Pérez, A.J., Labrador, M.A., Posada, J.D.: Centinela: a human activity recognition system based on acceleration and vital sign data. Pervasive Mobil. Comput. **8**(5), 717–729 (2012)

28. Rasna, M.J., Hossain , T., Inoue, S., Sha, S.S., Rahman, S., Ahad, M.A.R.: Supervised and neural classifiers for locomotion analysis. In: 2018 ACM International Joint Conference on Pervasive and Ubiquitous Computing and the 2018 International Symposium on Wearable Computers (UbiComp/ISWC) (2018)

29. Chetty, G., White, M., Akther, F.: Smart phone based data mining for human activity recognition. Proc. Comput. Sci. **46**, 1181–1187 (2015)

30. Feng, Z., Mo, L., Li, M.: A random forest-based ensemble method for activity recognition. In: 2015 37th Annual International Conference of the IEEE Engineering in Medicine and Biology Society (EMBC), pp. 5074–5077. IEEE (2015)

31. Catal, C., Tufekci, S., Pirmit, E., Kocabag, G.: On the use of ensemble of classifiers for accelerometer-based activity recognition. Appl. Soft Comput. **37**, 1018–1022 (2015)

32. Guo, H., Chen, L., Peng, L., Chen, G.: Wearable sensor based multimodal human activity recognition exploiting the diversity of classifier ensemble. In: Proceedings of the 2016 ACM International Joint Conference on Pervasive and Ubiquitous Computing, pp. 1112–1123 (2016)

33. Fatima, I., Fahim, M., Lee, Y-K., Lee, S.: Classifier ensemble optimization for human activity recognition in smart homes. In: Proceedings of the 7th International Conference on Ubiquitous Information Management and Communication, pp. 1–7 (2013)

34. Janidarmian, M., Fekr, A.R., Radecka, K., Zilic, Z.: A comprehensive analysis on wearable acceleration sensors in human activity recognition. Sensors **17**(3), 529 (2017)
35. Lee, J., Kim, J.: Energy-efficient real-time human activity recognition on smart mobile devices. Mobil. Inf. Syst. **2016**, (2016)
36. Yurur, O., Liu, C.H., Moreno, W.: A survey of context-aware middleware designs for human activity recognition. IEEE Commun. Mag. **52**(6), 24–31 (2014)
37. Torres-Huitzil, C., Alvarez-Landero, A.: Accelerometer-based human activity recognition in smartphones for healthcare services. In: Mobile Health, pp. 147–169. Springer (2015)
38. Xing, S., Tong, H., Ji, P.: Activity recognition with smartphone sensors. Tsinghua Sci. Technol. **19**(3), 235–249 (2014)
39. Nweke, H.F., Teh, Y.W., Al-Garadi, M.A., Alo, U.R.: Deep learning algorithms for human activity recognition using mobile and wearable sensor networks: State of the art and research challenges. Expert Syst. Appl. **105**, 233–261 (2018)
40. Zappi, P., Roggen, D., Farella, E., Tröster, G., Benini, L.: Network-level power-performance trade-off in wearable activity recognition: a dynamic sensor selection approach. ACM Trans. Embed. Comput. Syst. (TECS) **11**(3), 1–30 (2012)
41. Liang, Y., Zhou, X., Zhiwen, Y., Guo, B.: Energy-efficient motion related activity recognition on mobile devices for pervasive healthcare. Mobil. Netw. Appl. **19**(3), 303–317 (2014)
42. Basterretxea, K., Echanobe, J., Campo, I.: A wearable human activity recognition system on a chip. In: Proceedings of the 2014 Conference on Design and Architectures for Signal and Image Processing, pp. 1–8. IEEE (2014)
43. Islam, Z.Z., Tazwar, S.M., Islam, M.Z., Serikawa, S., Ahad, M.A.R. Automatic fall detection system of unsupervised elderly people using smartphone. In: Annual Conference on Artificial Intelligence. IEEE (2017)
44. Amiribesheli, M., Denmanoour, A., Bouchachia, A,: A review of smart homes in healthcare. J. Ambient Intel. Humaniz. Comput. **6**(4), 495–517 (2015)
45. Syeda, U.H., Zafar, Z., Islam, Z.Z., Tazwar, S.M., Rasna, M.J., Kise, K., Ahad, M.A.R.: Visual face scanning and emotion perception analysis between autistic and typically developing children. In: ACM UbiComp Workshop on Mental Health and Well-being: Sensing and Intervention. ACM (2017)
46. Saha, S.S., Rahman, S., Rasna, M.J., Zahid, T.B., Mahfuzul Islam, A.K.M., Ahad, M.A.R.: Feature extraction, performance analysis and system design using the du mobility dataset. IEEE Access **6**, 44776–44786 (2018)
47. Saha, S.S., Rahman, S., Rasna, M.J., Mahfuzul Islam, A.K.M., Ahad, M.A.R.: Du-md: an open-source human action dataset for ubiquitous wearable sensors. In: Joint 7th International Conference on Informatics, Electronics & Vision, 2nd International Conference on Imaging, Vision & Pattern Recognition (2018)
48. Ahmed, M., Antar, A.D., Ahad, M.A.R.: An approach to classify human activities in real-time from smartphone sensor data. In: 2019 Joint 8th International Conference on Informatics, Electronics Vision (ICIEV) and 2019 3rd International Conference on Imaging, Vision Pattern Recognition (icIVPR), pp. 140–145 (2019)
49. Liu, J., Yuan, Z., Wang, C.: Towards good practices for multi-modal fusion in large-scale video classification (2018). CoRR, arXiv:1809.05848
50. Celisse, A., et al.: Optimal cross-validation in density estimation with the $l^2$-loss. Ann. Stat. **42**(5), 1879–1910 (2014)

Printed in the United States
by Baker & Taylor Publisher Services